XL Symposium on Nuclear Physics (Cocoyoc 2017)

Journal of Physics: Conference Series Volume 876

Cocoyoc, Mexico
4 - 7 January 2017

Editors:

Luis Acosta
Roelof Bijker
Arturo Gomez-Camacho
Victor Velazquez

ISBN: 978-1-5108-4445-2
ISSN: 1742-6588

Printed from e-media with permission by:

Curran Associates, Inc.
57 Morehouse Lane
Red Hook, NY 12571

Some format issues inherent in the e-media version may also appear in this print version.

Printed by Curran Associates, Inc. (2017)

Institute of Physics
Dirac House, Temple Back
Bristol BS1 6BE UK

Phone: 44 1 17 929 7481
Fax: 44 1 17 920 0979

techtracking@iop.org

Additional copies of this publication are available from:

Curran Associates, Inc.
57 Morehouse Lane
Red Hook, NY 12571 USA
Phone: 845-758-0400
Fax: 845-758-2633
Email: curran@proceedings.com
Web: www.proceedings.com

Table of contents

Volume 876

XL Symposium on Nuclear Physics 2017 (Cocoyoc2017)
4–7 January 2017, Cocoyoc, Morelos, México

Accepted papers received: 7 July 2017
Published online: 14 July 2017

Preface

011001
OPEN ACCESS
XL Symposium on Nuclear Physics 2017 (Cocoyoc2017)

011002
OPEN ACCESS
XL Symposium on Nuclear Physics 2017 (Cocoyoc2017)

011003
OPEN ACCESS
Peer review statement

Papers

012001
OPEN ACCESS
Elastic scattering of $^{10}C + ^{27}Al$

E F Aguilera, J J Kolata, E Martinez-Quiroz, D Lizcano, P Amador-Valenzuela, A García-Flores, D W Bardayan, P D O'Malley, D S Monteiro, V. Morcelle, S. Carmichael, S L Henderson, D Blankstein, M R Hall, B Schultz, J Allen and J Kelly.....1

012002
OPEN ACCESS
Measurements of angular distributions for ^{7}Li elastically scattered from ^{58}Ni at energies around the Coulomb barrier

P Amador-Valenzuela, E F Aguilera, E Martinez-Quiroz, D Lizcano and J C Morales-Rivera.....9

012003
OPEN ACCESS
Overview of the data analysis and new micro-pattern gas detector development for the Active Target Time Projection Chamber (AT-TPC) project.

Yassid Ayyad, Wolfgang Mittig, Daniel Bazin and Marco Cortesi.....16

012004
OPEN ACCESS
Pentaquark states with hidden charm

Roelof Bijker.....22

012005
OPEN ACCESS
Neutron decay correlations in the Nab experiment

L. J. Broussard, R. Alarcon, S Baeßler, L. Barrón Palos, N. Birge, T. Bode, J. D. Bowman, T. Brunst, J. R. Calarco, J. Caylor, T. Chupp, V. Cianciolo, C. Crawford, G. W. Dodson, J. DuBois, W. Fan, W. Farrar, N. Fomin, E. Frlež, J. Fry, M. T. Gericke, F. Glück, G. L. Greene, R. K. Grzywacz, V. Gudkov, C. Hendrus, F. W. Hersman, T. Ito, H. Li, N. Macsai, M. F. Makela, J. Mammei, R. Mammei, J. Martin, M. Martinez, P. L. McGaughey, S. Mertens, J. Mirabal-Martinez, P. Mueller, S. A. Page, S. I. Penttilä, R. Picker, B. Plaster, D. Počanić, D. C. Radford, J. Ramsey, K. P. Rykaczewski, A. Salas-Bacci, E. M. Scott, S. K. L. Sjue, A. Smith, E. Smith, A. Sprow, E. Stevens, J. Wexler, R. Whitehead, W. S. Wilburn, A. R. Young and B. A. Zeck.....28

012006
OPEN ACCESS
Investigation of the Hoyle state in ^{12}C with a new hodoscope detector

D Dell'Aquila, I Lombardo, M Vigilante, M De Luca, L Acosta, C Agodi, F Cappuzzello, D Carbone, M Cavallaro, S Cherubini, A Cvetinovic, G D'Agata, L Francalanza, G L Guardo, M Gulino, I Indelicato, M La Cognata, L Lamia, A Ordine, R Pizzone, S Puglia, G Rapisarda, S Romano, G Santagati, R Spartà, C Spitaleri, A Tumino and G Verde.....38

012007
OPEN ACCESS
Pentaquark photoproduction

César Fernández-Ramírez, Astrid N. Hiller Bin and Alessandro Pilloni.....44

012008
OPEN ACCESS
Vibrations and potential energy surfaces (with Argonne V18) of ^{4}He and ^{3}He

Lorenzo Fortunato.....48

012009
OPEN ACCESS
Current and Future Tests of the Algebraic Cluster Model of ^{12}C

Moshe Gai.....56

012010
OPEN ACCESS
The baryon spectroscopy: strong decays and strange suppression*

H. García-Tecocoatzi.....61

012011
OPEN ACCESS
Effect of breakup and resonance states of ^{6}Li on fusion

A. Gómez Camacho.....67

012012
OPEN ACCESS
In search of a broader microscopic underpinning of the potential energy surface in heavy deformed nuclei

P. O. Hess and M. Ermamatov.....77

012013
OPEN ACCESS
On the investigation of resonances above and below the threshold in nuclear reactions of astrophysical interest using the Trojan Horse Method.

M. La Cognata, C. Spitaleri, S. Cherubini, M. Gulino, I. Indelicato, L. Lamia, R.G. Pizzone, S. Romano, O. Trippella and A. Tumino.....85

012014
OPEN ACCESS
Confinement in a correlated Instanton-Dyon Ensemble

Miguel Angel Lopez-Ruiz, Yin Jiang and Jinfeng Liao.....93

012015
OPEN ACCESS
Study of the ^{12}C excited states above the Hoyle State.

E. López-Saavedra, L. Acosta, V. Araujo, F. Favela, A. Huerta, J. Aspiazu, G. Murillo, R. Policroniades, P. Santa Rita, A. Varela and E. Chávez.....100

012016
OPEN ACCESS
Main Nuclear Physics requirements for the robustness of *r*-process nucleosynthesis calculations in slow ejecta from neutron-star mergers (NSM)

J J Mendoza-Temis and A Frank.....109

012017
OPEN ACCESS
Towards the measurement of the $^{13}C(d, p)^{14}C$ cross section using AMS

S. Murillo-Morales, L. Barrón-Palos, E. Chávez and V. Araujo-Escalona.....119

012018
OPEN ACCESS
Aluminum and Oxygen Isotopic Ratios in Meteorite Grains: a Puzzle Solved By Nuclear and Stellar Physics

Sara Palmerini.....124

012019
OPEN ACCESS
Global Correlations for Low-Lying Collective 2^+ States

Z. Z. Qin, Y. Lei, S. Pittel and R. Bijker.....131

012020
OPEN ACCESS
The theory of hadronic parity violation

Matthias R Schindler.....141

012021
OPEN ACCESS
Oxygen-15+α resonant elastic scattering to study cluster states in ^{19}Ne

D Torresi, C. Wheldon, Tz. Kokalova, S. Bailey, A. Boiano, C. Boiano, M. Cavallaro, S. Cherubini, A. Di Pietro, J.P. Fernandez Garcia, M. Fisichella, T.R. Glodariu, J. Grebosz, M. La Cognata, M. La Commara, M. Lattuada, M. Mazzocco, D. Mengoni, C. Parascandolo, D. Pierroutsakou, G. Pizzone, C. Signorini, C. Stefanini, L. Stroe, C. Spitaleri, E. Strano and M. Zadro.....148

012022
OPEN ACCESS
TDA and RPA pseudoscalar and vector solutions for the low energy regime of a motivated QCD Hamiltonian.

T. Yépez-Martínez, D. A. Amor Quiroz, P. O. Hess and O. Civitarese.....155

012023
OPEN ACCESS
Muon study for gamma/hadron air-shower discrimination in the HAWC observatory

A. Zuñiga Reyes, A. Hernández Almada, A. Sandoval Espinosa the HAWC Collaboration.....163

IOP Conf. Series: Journal of Physics: Conf. Series **876** (2017) 011001 doi:10.1088/1742-6596/876/1/011001

XL Symposium on Nuclear Physics

Hacienda Cocoyoc, Morelos, México
January 4-7, 2017

Editors:

Luis Acosta
Instituto de Física
Universidad Nacional Autónoma de México
Roelof Bijker
Instituto de Ciencias Nucleares
Universidad Nacional Autónoma de México
Arturo Gómez-Camacho
Instituto Nacional de Investigaciones Nucleares
Víctor Velázquez
Facultad de Ciencias
Universidad Nacional Autónoma de México

IOP Conf. Series: Journal of Physics: Conf. Series **876** (2017) 011001 doi:10.1088/1742-6596/876/1/011001

International Advisory Committee

Osvaldo Civitarese	Universidad Nacional de La Plata, Argentina
Daniel Bardayan	University of Notre Dame, USA
Francesco Cappuzzello	INFN-Laboratori Nazionali del Sud, Italia
Jerry Draayer	Louisiana State University, USA
Alfredo Galindo-Uribarri	Oak Ridge National Laboratory, USA
María José García Borge	ISOLDE-CERN, Switzerland
Reiner Krücken	TRIUMF, Canada
Jorge López	University of Texas at El Paso, USA
Takashi Nakamura	Tokyo Institute of Technology, Japan
Thomas Papenbrock	University of Tennessee, USA
Jorge Piekarewicz	Florida State University, USA
Stuart Pittel	University of Delaware, USA
Elena Santopinto	Università degli Studi di Genova, Italia
W. Michael Snow	Indiana University, USA
Adam Szczepaniak	Indiana University, USA
Piet Van Isacker	GANIL, France
Michael Wiescher	University of Notre Dame, USA

Organizing Commmittee

Luis Acosta	IF-UNAM
Roelof Bijker	ICN-UNAM
Arturo Gómez-Camacho	ININ
Víctor Velázquez	FC-UNAM

Sponsors

Instituto de Ciencias Nucleares, UNAM
Instituto de Física, UNAM
Instituto Nacional de Investigaciones Nucleares
División de Física Nuclear de la SMF

IOP Conf. Series: Journal of Physics: Conf. Series **876** (2017) 011001 doi:10.1088/1742-6596/876/1/011001

Photograph of the event

IOP Conf. Series: Journal of Physics: Conf. Series **876** (2017) 011001 doi:10.1088/1742-6596/876/1/011001

Preface

The XL Symposium on Nuclear Physics was celebrated at the Hacienda Cocoyoc from January 4-7, 2017. The Symposium was conceived from the beginning as a small meeting designed to bring together some of the leading nuclear scientists in the field. Its most distinctive feature is to provide a forum for specialists in different areas of nuclear physics, both theorists and experimentalists, students, postdocs and senior scientists, in a relaxed and informal environment providing them with a unique opportunity to exchange ideas. After the first meeting in Oaxtepec in 1978, the Symposium was organized every year without interruption, which makes the present one the 40th in a row.

In celebration of the 40th anniversary, at the beginning of each session there were special presentations to reflect on the past, present and future of this series of symposia by our distinguished colleagues, Alejandro Frank, Elí Aguilera, Peter Hess, María Esther Ortiz and Stuart Pittel, all of whom have participated in the symposium since the early days. They shared many memories, photographs, anecdotes and stories on different aspects of the symposia, such as the history, the location, the people, the program and the proceedings, providing a nice mixture of emotions, nostalgia and reflection.

The scientific program consisted of 32 invited talks and 16 posters on a wide variety of hot topics in contemporary nuclear physics, ranging from the traditional fields of nuclear structure and reactions to nuclear astrophysics, hadronic physics, fundamental symmetries and relativistic heavy ions. At the meeting there were representatives of many large international collaborations to report on the most recent progress, such as the HAWC Observatory in Mexico, FRIB in the USA, and the ALICE and ATLAS Collaborations at CERN, Switzerland. In addition, there were talks and poster presentations on the most recent developments in other laboratories like Notre Dame, RIKEN, RCNP, JLab, Oak Ridge, INFN-Legnaro, INFN-LNS, GANIL and TRIUMF, as well as the results obtained in local Mexican facilities like ININ, LEMA and the Van de Graaff Laboratory Carlos Graef at IF-UNAM. On the theoretical side, interesting new developments were presented in nuclear structure, weakly bound nuclei, nuclear clusters, shell model, hadronic physics and low-energy QCD.

The high quality of the talks, the prestige of the speakers and the broad spectrum of subjects covered in the meeting, shows that nuclear physics is a very active area at the frontier of scientific research which establishes bridges

IOP Conf. Series: Journal of Physics: Conf. Series **876** (2017) 011001 doi:10.1088/1742-6596/876/1/011001

between many different disciplines. This special volume of Journal of Physics: Conference Series contains 23 contributions that were presented at this year's symposium.

Finally, we would like to thank all our colleagues for their enthusiastic participation in this years meeting. In particular, we are grateful to the members of the International Advisory Committee for their invaluable suggestions and to the anonymous referees for their time and dedication in helping us to maintain the high standard of the contributions published in the proceedings of the Symposium.

Luis Acosta
Roelof Bijker
Arturo Gómez-Camacho
Víctor Velázquez

First Author	Title
Aguilera	Elastic scattering of ^{10}C + ^{27}Al
Amador-Valenzuela	Measurements of angular distributions for ^{7}Li elastically scattered from ^{58}Ni at energies around the Coulomb barrier
Ayyad	Overview of the data analysis and new micro-pattern gas detector development for the Active Target Time Projection Chamber (AT-TPC) project
Bijker	Pentaquark states with hidden charm
Broussard	Neutron decay correlations in the Nab experiment
Dell'Aquila	Investigation of the Hoyle state in ^{12}C with a new hodoscope detector
Fernández-Ramírez	Pentaquark photoproduction
Fortunato	Vibrations and potential energy surfaces (with Argonne V18) of ^{4}He and ^{3}He
Gai	Current and Future Tests of the Algebraic Cluster Model of ^{12}C
García-Tecocoatzi	The baryon spectroscopy: strong decays and strange suppression
Gómez-Camacho	Effect of breakup and resonance states of ^{6}Li on fusion
Hess	In search of a broader microscopic underpinning of the potential energy surface in heavy deformed nuclei
La Cognata	On the investigation of resonances above and below the threshold in nuclear reactions of astrophysical interest using the Trojan Horse Method
Lopez-Ruiz	Confinement in a correlated Instanton-Dyon Ensemble
López-Saavedra	Study of the ^{12}C excited states above the Hoyle State
Mendoza-Temis	Main Nuclear Physics requirements for the robustness of r-process nucleosynthesis calculations in slow ejecta from neutron-star mergers (NSM)
Murillo-Morales	Towards the measurement of the $^{13}\mathrm{C}(d,p)^{14}\mathrm{C}$ cross section using AMS
Palmerini	Aluminum and Oxygen Isotopic Ratios in Meteorite Grains: a Puzzle Solved by Nuclear and Stellar Physics
Pittel	Global Correlations for Low-Lying Collective 2^+ States
Schindler	The theory of hadronic parity violation
Torresi	$^{15}\mathrm{O} + \alpha$ resonant elastic scattering to study cluster states in ^{19}Ne
Yépez-Martínez	TDA and RPA pseudoscalar and vector solutions for the low-energy regime of a motivated QCD Hamiltonian
Zúñiga-Reyes	Muon study for gamma/hadron air-shower discrimination in the HAWC observatory

IOP Conf. Series: Journal of Physics: Conf. Series **876** (2017) 011003 doi:10.1088/1742-6596/876/1/011003

Peer review statement

All papers published in this volume of *Journal of Physics: Conference Series* have been peer reviewed through processes administered by the proceedings Editors. Reviews were conducted by expert referees to the professional and scientific standards expected of a proceedings journal published by IOP Publishing.

Elastic scattering of ^{10}C + ^{27}Al

E F Aguilera[1,*], J J Kolata[2], E Martinez-Quiroz[1], D Lizcano[1], P Amador-Valenzuela[1], A García-Flores[1,3], D W Bardayan[2], P D O'Malley[2], D S Monteiro[2,5], V. Morcelle[4], S. Carmichael[2], S L Henderson[2], D Blankstein[2], M R Hall[2], B Schultz[2], J Allen[2], J Kelly[2]

[1] Departamento de Aceleradores, Instituto Nacional de Investigaciones Nucleares, Apartado Postal 18-1027, C. P. 11801, México, D. F., México
[2] Physics Department, University of Notre Dame, Notre Dame, In, 46556-5670, USA
[3] Universidad Autónoma del Estado de México, Código Postal 50000, Toluca, México
[4] Departamento de Física, Universidade Federal Rural do Rio de Janeiro, 23851-970, Rio de Janeiro, Brasil
[5] Instituto Latino-Americano de Ciências da Vida e da Natureza - UNILA - Av. Tancredo Neves, 6731, Bl. 6. CEP 85867- 970 - Foz do Iguaçu, Paraná - Brasil

E-mail: *eli.aguilera@inin.gob.mx

Abstract. Preliminary results for the elastic scattering of the Super-Borromean nucleus ^{10}C on ^{27}Al are presented, at E_{lab} = 29 MeV. Taking the Sao Paulo potential as the bare potential, possible polarization potentials are investigated by adding a volume and a surface complex term to account for fusion and direct couplings, respectively. Besides an imaginary short-range potential to simulate the bulk of fusion, the best-fit polarization potential turns out to be real and has the effect of shifting the barrier radius up by 0.6 fm, thus suggesting a fusion enhancement.

1. Introduction

There has been considerable interest in studying Borromean nuclei, which can be described as nuclei having a three-cluster structure with the additional property that the removal of any of the three clusters leads to an unbound two-body system. The name "Borromean" is borrowed from the so called Borromean rings in mathematics, which consist of three topological circles which are linked in such a way that removing any ring results in two unlinked rings. Their use in the coat of arms of the aristocratic Borromeo family in Northern Italy (15th century) suggested the name.

There is also one known case of a four-cluster nucleus with Borromean properties, the ^{10}C nucleus, which can be described as two alpha particles plus two protons. In this case, the removal of any one of the four clusters produces an unbound three-cluster system, which has led some authors to call this a Super-Borromean nucleus [1]. More precisely, this is an example of a 4th order Brunnian link [2].

With the aim of investigating the behaviour of this exotic nucleus in interactions with light heavy-ions, an elastic scattering angular distribution has been measured for the ^{10}C + ^{27}Al system for an energy E_{lab} = 29 MeV. Preliminary results of these measurements are shown in the present work, as well as the results of corresponding optical model analyses.

IOP Conf. Series: Journal of Physics: Conf. Series **876** (2017) 012001 doi:10.1088/1742-6596/876/1/012001

The experimental procedure is described in Sec. 2) as well as the preliminary results and a comparison with previous data for a similar system. An appropriate optical model analysis is described in Sec. 3 and, finally, a summary and the main conclusions of this work will be presented in Section 4.

2. Experimental procedure and results

A primary 42 MeV ^{10}B beam was obtained from the FN Tandem accelerator at the University of Notre Dame (UND). This beam impinged on a ^{3}He primary target which was contained into a gas cell held at 1 atm pressure. The secondary ^{10}C beam was thus obtained through the n-p exchange reaction ^{3}He(^{10}B, ^{10}C). In addition to ^{10}C, satellite beams of ^{10}B and ^{11}C were also produced, with smaller yields of lighter beams such as Be and Li isotopes. In order to separate between reactions induced by the different secondary projectiles, the beam was bunched and the respective time-of-flight was measured. After bunching, the typical yields were 40 particle-nA for the primary beam and 3.9×10^{4} ^{10}C/sec for the secondary beam.

Three E-ΔE telescopes made of Silicon surface barrier (SSB) detectors were used at the most forward angles, while three additional single SSB detectors were used at larger angles. The most forward telescope had a 7 mm collimator while the diameter of the remaining detectors varied between 10 and 22 mm. Figure 1 shows the experimental setup. The collimator-target distance was 15 cm for the left-hand detectors and 18 cm for the right-hand ones. The target was an Al foil with thickness 158.7 μg/cm^2. A 2.59 mg/cm^2 Au target was also used for normalization purposes.

Figure 1. Experimental setup. As indicated, three SSB E-ΔE telescopes were used while the remaining three were single SSB detectors.

The plate holding the detectors (Fig. 1) was rotated twice so that three different angular configurations were measured, covering the angular range $15^\circ \leq \theta_{lab} \leq 65^\circ$. Figures 2 and 3 show examples of the typical ΔE vs E and T vs E spectra illustrating the achieved separation of the elastic ^{10}C groups.

Figure 4 presents the obtained angular distribution for ^{10}C on ^{27}Al (circles). For comparison, data previously reported [3] for the neighboring (normal) system ^{12}C + ^{27}Al (squares) are also shown, which correspond to a similar energy in the center of mass reference frame ($E_{c.m.} \sim 21$ MeV).

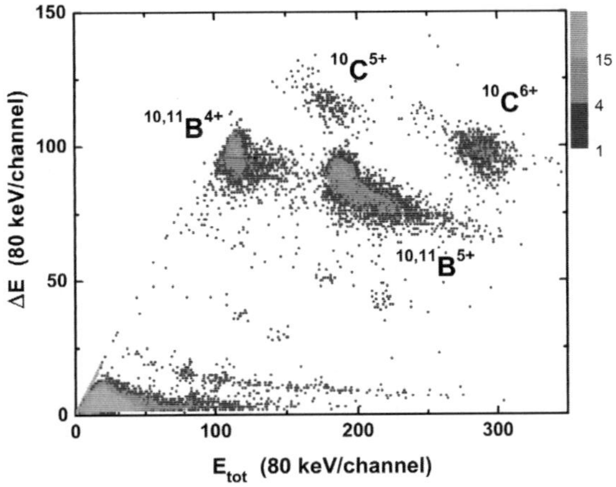

Figure 2. Typical ΔE vs E spectrum, taken at $E_{lab} = 29.5$ MeV and $\theta_{lab} = 30°$.

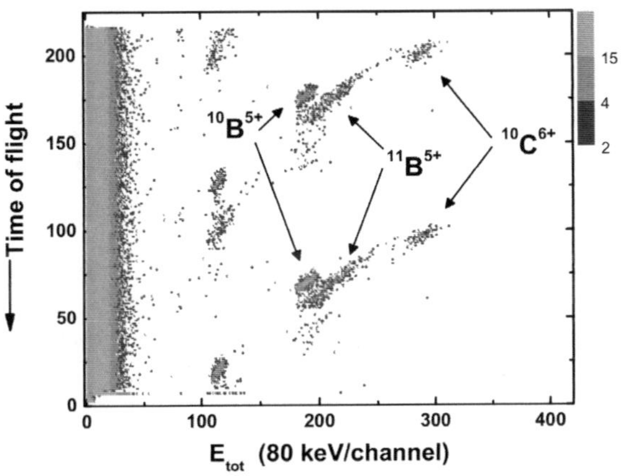

Figure 3. Typical Time-Of-Flight vs Energy spectrum, taken at $E_{lab} = 29.5$ MeV and $\theta_{lab} = 30°$.

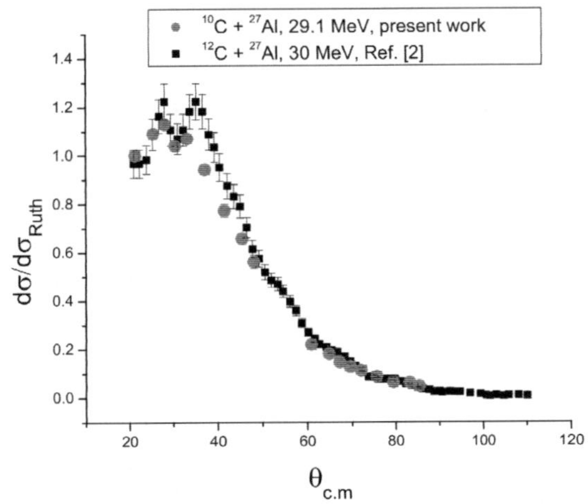

Figure 4. Experimental angular distribution obtained for the ^{10}C on ^{27}Al system at $E_{lab} = 29$ MeV. Both statistical uncertainties and estimated angle uncertainties have been included. For comparison, earlier data [3] reported for the stable ^{12}C on ^{27}Al system at nearly the same energy in the center of mass reference frame are also shown.

The structure shown by the ^{12}C data around 30° is also insinuated in the ^{10}C results, but much less pronounced. In addition, relative to the ^{12}C data, the angular distribution for the exotic system seems to be damped for angles in the region $35° \leq \theta_{c.m.} \leq 45°$. At the most backward angles, both angular distributions coincide. Because of the mass difference between $^{10,12}C$, it should be more appropriate to transform the angle into distance of closest approch (see Sect. 3.2) to make the comparison. This is done in Fig. 5.

Qualitatively, the above observations remain valid if one replaces 30° by 2.5 fm and " 35° $\leq \theta_{c.m.} \leq$ 45° " by " 1.9 fm $\leq d \leq$ 2.3 fm " in the above paragraph. In next section, an optical model analysis will be performed for ^{10}C + ^{27}Al.

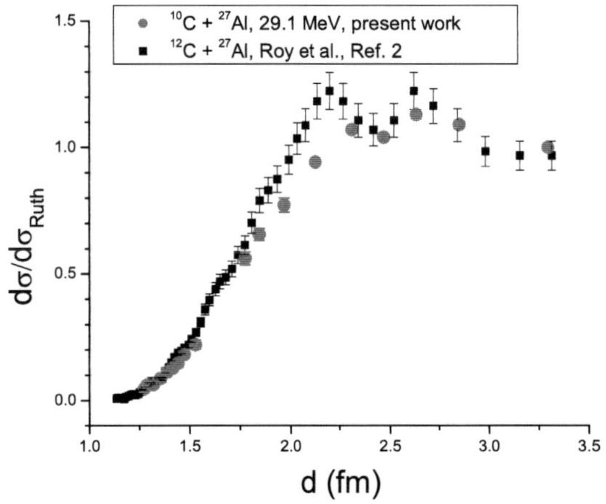

Figure 5. Same as Fig. 4, but $\theta_{c.m.}$ has been mapped into the distance of closest approach (see Sect. 3.2).

3. Optical model analysis

It's been long known that the parameter-free São Paulo Potential (SPP) [4] can be used as an appropriate bare potential to describe fusion reactions. Crema *et al.* [5], for example, have shown that the SPP gives reliable results in the analysis of fusion reactions with stable weakly-bound projectiles. Among other results, this fact has been used to show that the experimental fusion excitation functions for neutron-halo projectiles are suppressed with respect to the corresponding SPP predictions for energies above the respective Coulomb barriers, while those for proton-halo systems are enhanced [7,8].

Additional evidence exists that at least for another proton-rich nucleus, ^{7}Be, a fusion enhancement above the barrier can be observed [8,9], albeit weaker than that in the proton-halo case. Since ^{10}C is also a proton-rich nucleus, it would be interesting to investigate whether a fusion enhancement might be expected from its interaction with other nuclei. Fusion data have not been measured for the present system, but an optical model analysis of the elastic scattering data might shed some light on this issue.

Within the framework of the optical model, fusion has been described as the absorption in a short-range imaginary potential interior to the barrier, W_{int}. Such a potential effectively simulates an incoming-wave boundary condition, thus giving a good barrier penetration model (BPM) prediction. In order for the fusion cross section calculated this way to be realistic, the optical model wave function should correctly describe the respective elastic scattering angular distribution. So, an appropriate polarization potential should be usually added to the bare potential. Assumedly, this polarization potential would result from respective couplings to fusion and direct channels. It has been shown for many systems that it is reasonable to decompose the total polarization potential into a direct and a fusion part [10–18]. With this in mind, the following general form for the total optical potential is proposed:

$$U_{TOT} = SPP + V_{Coul} + W_{int} + U_F + U_S, \tag{1}$$

where SPP is the bare (Sao Paulo) potential, V_{Coul} is the Coulomb potential, W_{int} is a short range Woods-Saxon potential (interior to the barrier: W_0 = 50 MeV, r_0 = 1. fm, a_0 = 0.2 fm), and U_F, U_S are complex polarization potentials that can be associated to couplings to the fusion and direct channels, respectively. U_F (U_S) is taken as having a volume (derivative) Woods-Saxon shape, where in each case the radius and difussenes are the same for both, the respective real and imaginary parts. To be more specific, $U_{F,S}$ can be written as

$$U_{F,S} = V_{F,S} + iW_{F,S}, \tag{2}$$

where V_F (V_S) and W_F (W_S) have identical radius and diffuseness.

3.1. Case of no polarization potential

The code FRESCO and its search version, SFRESCO [19], were used to perform all optical model calculations. As a reference, a calculation using SPP for the real part and 0.78×SPP for the imaginary part was performed. This combination has been shown to provide a good description of many systems [4]. For the present case, the result is illustrated with the dashed line in Fig. 6, which evidently does not follow the data. This indicates that the present system falls out of the systematics and should be dealt with separately.

The case where both U_F and U_S are zero was also calculated and it is shown with the dotted line in Fig. 6 (label SPP + W$_{int}$). Clearly, a non-zero polarization potential is necessary to describe the data.

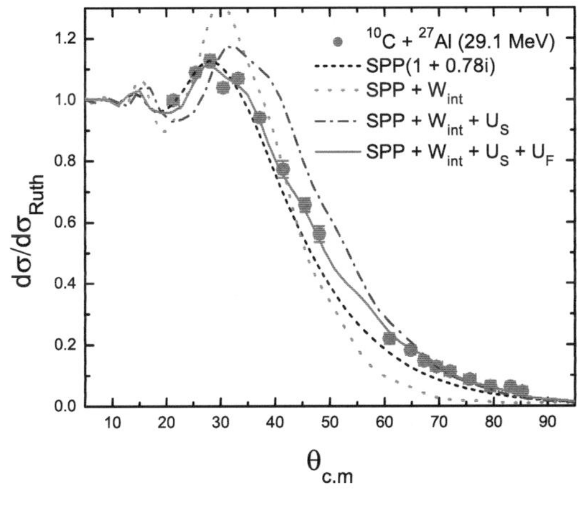

Figure 6. Optical model calculations, as explained in the text.

3.2. Purely direct polarization potential

In a first trial, the fusion part U_F in eq. 1 is assumed to be zero. As a guide to choose values for the geometric parameters r$_S$, a$_S$ (reduced radius and diffuseness, respectively) of the Woods-Saxon potential whose derivative will give U_S, a transformation $\theta_{c.m.} \rightarrow d$ is performed on the data. In this transformation, d is the reduced distance of closest approach on a Rutherford trajectory, given by

$$D = d(A_p^{1/3} + A_t^{1/3}) = \frac{1}{2}D_0\left(1 + \frac{1}{sin(\theta_{c.m.}/2)}\right), \quad with\ D_0 = \frac{Z_p Z_t e^2}{E_{c.m.}}. \tag{3}$$

The transformed data are shown in Fig. 7, where the solid curve represents a fit of the data with the function

$$f(x) = \frac{a}{1 + exp[-k(x - x_c)]}, \tag{4}$$

where a, k and x_c are free parameters whose best-fit values are given in the inset of the figure. The strong absorption distance, d_S, corresponds to the point where the ratio $\mathrm{d}\sigma/\mathrm{d}\sigma_{Ruth} = 0.25$ and is calculated from $f(x)$ as $\mathrm{d}_S = 1.55$ fm.

If one assumes that the falling down of the data with decreasing d is due only to the surface potential U_S, it is reasonable to take the radius of this potential as $\mathrm{r}_S = \mathrm{d}_S = 1.55$ fm, with an appropriate diffuseness to reproduce the slope of $f(d)$ (solid curve), such as does the dotted curve in Fig. 7. This gives $a_S = 0.9$ fm. The strength given in the inset ($\mathrm{W}_{S0} = 0.666$) has no physical meaning and was chosen (along with a_S) to match the right wing of the dotted curve to the solid curve. The actual strengths of U_S (V_{S0}, W_{S0}) will be obtained from fitting the experimental angular distribution.

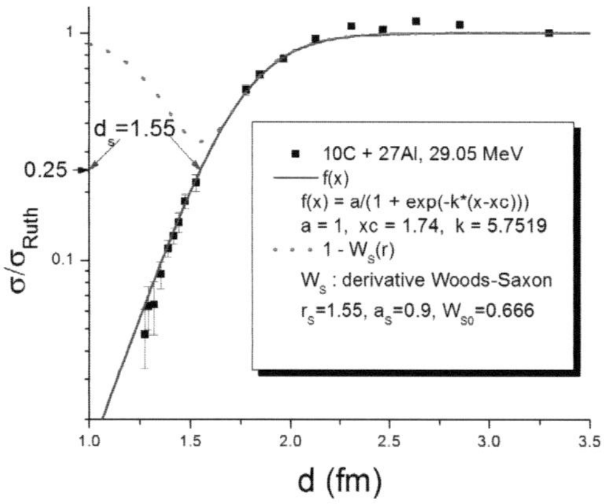

Figure 7. Plot of $\mathrm{d}\sigma/\mathrm{d}\sigma_{Ruth}$ vs d for the ^{10}C + ^{27}Al system. The meaning of the curves is explained in the inset and in the text.

The best fit is obtained for V_{S0} = -0.7 MeV, $W_{S0} = 0$, but a good description of the data is not achieved ($\chi^2/\mathrm{N} = 38$), as shown by the dash-dotted line in Fig. 6 (label SPP + W_{int} +U_S). This probably means that the falling down of the data with decreasing d should be produced by a combination of a fusion and a direct polarization potential.

3.3. Fusion and direct polarization potentials

The fusion polarization potential U_F will now be included, with radius and diffuseness values of $\mathrm{r}_F = 1.4$ fm, $\mathrm{a}_F = 0.43$ fm. These values have been used in similar descriptions of other systems, for instance the ^{6}Li + ^{208}Pb system [12]. Keeping these values fixed, the four strengths V_{F0}, W_{F0}, V_{S0}, W_{S0} were varied. The radius and diffuseness of the direct part, r_S, a_S were also varied, but with the constraint that their sum should be fixed at the value derived in the previous section, *i.e.*, $\mathrm{r}_S + \mathrm{a}_S = 1.55 + 0.9 = 2.45$ fm. This constraint would in principle guarantee the intuitive expectation that the shape of the surface potential should be able to partially match the solid curve in Fig. 7. This procedure yielded the best-fit values V_{F0} = 14.87 MeV, W_{F0} = 0, V_{S0} = -0.53 MeV, $W_{S0} = 0$, $\mathrm{r}_S = 1.77$ fm, $\mathrm{a}_S = 0.68$ fm. Varying the value of r_F around 1.4 fm did not improve the fit, which yielded $\chi^2/\mathrm{N} = 2.8$ as a final value. The respective results are represented by the solid curve in Fig. 6 (label SPP +W_{int} +U_S + U_F).

Notice that null absorptive potentials are obtained for both the fusion and the direct parts, with real polarization potentials which are attractive (repulsive) for the fusion (direct) part. This is illustrated in Fig. 8. Comparing the bare (solid curve) with the total (dashed curve) potential, one concludes that the polarization potentials have the effect of shifting the barrier radius up by 0.63 fm, while the barrier height remains nearly unchanged. Since the fusion cross section varies roughly as the square of the barrier radius, a fusion enhancement is predicted by these results. The fact that W_{int} is the only imaginary part in the final potentials appearing in eq. 1, does actually imply that the fusion cross section would saturate the total reaction cross section at the present energy, which according to the present calculations amounts to σ_R (or σ_{Fus}) = 976 mb. Further experimental data are needed to corroborate the validity of these conclusions.

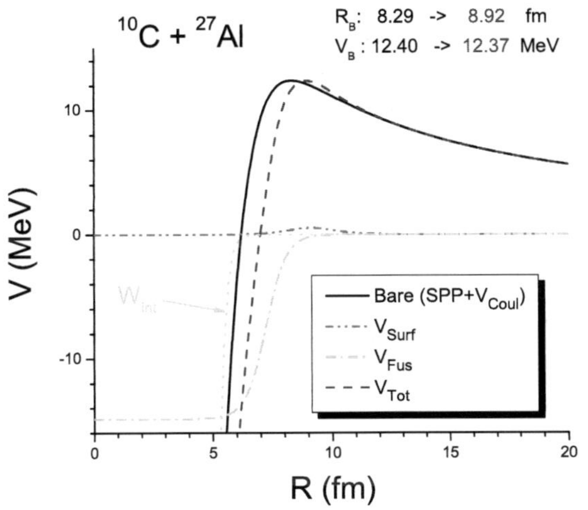

Figure 8. Different potentials involved in the OMP calculations. The final results for the direct, the fusion, and the total real potentials are represented by dash-double dot, dash-dot, and dashed curves, respectively. The bare (SPP) plus Coulomb potential is shown with the solid line, while the imaginary interior potential W_{int} is displayed with dots.

4. Conclusions

Preliminary data for the elastic scattering angular distribution of the ^{10}C + ^{27}Al system are presented, corresponding to $E_{c.m.}$ = 21.2 MeV (1.7×V_B). Comparison with reported data for the system ^{12}C + ^{27}Al at the same energy in the center of mass reference frame, indicates substantial differences. A purely surface (direct) polarization potential can not describe the data but introducing in addition a volume (fusion) potential, a good description is achieved. The best-fit polarization potential is non-absorptive, with the fusion (direct) part providing an attractive (repulsive) term. Altogether, the net effect is to shift the barrier radius up by 0.6 fm, thus predicting a fusion enhancement above the barrier. It would be interesting to corroborate this prediction with corresponding fusion measurements.

Acknowledgments

This work has been partially supported by CONACYT (México) and by NSF (USA) under grant No. PHY14-01343.

References

[1] Curtis N, Achouri N, Ashwood N, Bohlen H, Catford W, Clarke N, Freer M, Haigh P, Laurent B and Orr N 2008 *J. Phys.: Conf. Series* **111** 012022
[2] Brunn H 1892 Über Verkettung, S.-B.Math.-Phys.KI. Bayer Akad. Wiss. **22** 77
[3] Roy A, Frawley A D and Kemper K W 1979 *Phys. Rev.* C **20** 2143
[4] Chamon L C, Carlson B V, Gasques L R, Pereira D, De Conti C, Alvarez M A G, Hussein M S, Cândido Ribeiro M A, Rossi Jr. E S and Silva C P 2002 *Phys. Rev.* C **66** 014610
[5] Crema E, Chamon L C and Gomes P R S 2005 *Phys. Rev.* C **72** 034610
[6] Crema E, Gomes P R S and Chamon L C 2007 *Phys. Rev.* C **75** 037601
[7] Aguilera E F 2016 *J. Phys.: Conf. Series* **730** 012002

IOP Conf. Series: Journal of Physics: Conf. Series **876** (2017) 012001 doi:10.1088/1742-6596/876/1/012001

[8] Kolata J J ,Guimarães V and Aguilera E F 2016 *Eur. Phys. J.* A **52** 123
[9] Martinez-Quiroz E, Aguilera E F, Lizcano D, Amador-Valenzuela P, García-Martínez H, Kolata J J, Roberts A, Lamm L O, Rogachev G, Guimarães V, Becchetti F D, Villano A, Ojaruega M, Febbraro M, Chen Y, Jiang H, DeYoung P A and Peaslee G F 2014 *Phys. Rev.* C **90** 014616
[10] Kim B T, Naito M and Udagawa T 1990 *Phys. Lett.* B **237** 19
[11] Kim B T, So W Y, Hong S W and Udagawa T 2002 *Phys. Rev.* C **65** 044616
[12] So W Y, Udagawa T, Kim K S, Hong S W and Kim B T 2007 *Phys. Rev.* C **75** 024610
[13] Gómez-Camacho A and Aguilera E F 2004 *Rev. Mex. Fis.* **50** Supl. 3, 265
[14] Gómez-Camacho A and Aguilera E F 2004 *Nucl. Phys.* A **735** 425
[15] Gómez-Camacho A, Aguilera E F, Gomes P R S , Lubian J and Padron I 2007 *Nucl. Phys.* A **787** 275c
[16] Gómez-Camacho A, Gomes P R S, Lubian J, Aguilera E F and Padron I 2007 *Phys. Rev.* C **76** 044609
[17] Gómez-Camacho A, Aguilera E F, Lubian J and Gomes P R S 2013 *J. Phys. G: Nucl. Part. Phys.* **40** 035103
[18] Gómez-Camacho A and Aguilera E F 2014 *Phys. Rev.* C **90** 064607
[19] Thompson I 1988 *Comp. Phys. Rep.* **7** 167

IOP Conf. Series: Journal of Physics: Conf. Series **876** (2017) 012002 doi:10.1088/1742-6596/876/1/012002

Measurements of angular distributions for ^{7}Li elastically scattered from ^{58}Ni at energies around the Coulomb barrier

P Amador-Valenzuela[1], E F Aguilera[1], E Martinez-Quiroz[1], D Lizcano[1] and J C Morales-Rivera[1,2]

[1] Departamento de Aceleradores, Instituto Nacional de Investigaciones Nucleares, Apartado Postal 18-1027, C.P. 11801, Ciudad de México, México

[2] Facultad de Ciencias, Universidad Autónoma del Estado de México, C.P. 50000, Toluca, Estado de México

E-mail: paulina.amador@inin.gob.mx, eli.aguilera@inin.gob.mx

Abstract. Recently, experimental measurements of elastic scattering angular distributions for the system ^{7}Li+^{58}Ni at ten different energies around the Coulomb barrier were made by the Heavy-Ion Group. The measurements were made at the Tandem Van de Graaff Particle Accelerator Laboratory in the National Institute for Nuclear Research (ININ) in Mexico. In this work, preliminary elastic scattering angular distributions for five energies (E_{lab} = 12.0, 12.5, 13.0, 13.5 and 14.22 MeV) are presented. The preliminary experimental data were analyzed using the São Paulo Optical Model Potential (SPP) which is based on a double-folding potential, reproducing very well these data. A comparison is made with old data reported back in 1973 and in 2012. Further analysis is in progress in order to fully understand this particular system, specially because ^{7}Li is known to be a weakly bound nucleus.

1. Introduction

In nuclear physics, elastic scattering experiments are of great importance since valuable information on the properties of the nucleus can be obtained through them. A clear and specific example of the relevance of elastic scattering experiments is the discovery of the atomic nucleus in 1911, made by Ernest Rutherford. Lately, many experiments and theoretical analyse have been performed using radioactive beams as projectiles to produce a nuclear reaction [1, 2, 3, 4, 5, 6, 7, 8, 9, 10, 11, 12, 13, 14, 15, 16]. It is well known that in general experiments with halo nuclei are difficult to carry out, mainly because of the low intensities of the respective beams. It is also known that halo nuclei (cluster-type nuclei) have a relative small separation energy which means that the probability of breakup is important in the presence of the force field produced by the target. With this in mind, information on the effect of coupling to the breakup channel becomes essential when dealing with radioactive beams of this kind. This effect can first be investigated using stable weakly bound beams which have much higher intensities than radioactive beams. The suitable stable nuclei for this kind of study are ^{9}Be, ^{6}Li and ^{7}Li with separation energies of 1.57, 1.47 and 2.47 MeV, respectively.

Recently, many experiments of elastic scattering between heavy ions, using strongly bound nuclei, at energies near the Coulomb barrier have been performed. The analysis of these

IOP Conf. Series: Journal of Physics: Conf. Series **876** (2017) 012002 doi:10.1088/1742-6596/876/1/012002

experiments present a systematic behaviour of the energy dependence of the interacting optical potential known as the Threshold Anomaly (TA) [17]. Basically, the threshold anomaly implies a rapid variation of the optical potential (real and imaginary part) extracted from the analysis of the experimental data when the energy decreases towards the Coulomb barrier. For the real part of the potential there is a localized peak (bump) at the barrier energy, whereas there is a sharp decrease of the imaginary part of this potential. The decrease of the imaginary potential can be understood as the closure of the non-elastic channels at energies below the Coulomb barrier [18, 19]. For weakly bound nuclei the breakup channel is expected to be a relevant reaction channel even at energies below the Coulomb barrier. Since the coupling to the breakup channel produces a repulsive polarization potential then there is a big probability that the threshold anomaly may dissapear. It has been suggested that a new type of anomaly is present in the scattering of weakly bound nuclei, named as breakup threshold anomaly [20]. In this anomaly the strength of the imaginary potential increases as the incident energy decreases, at energies below the Coulomb barrier.

For ^{6}Li projectiles, it has been shown that the threshold anomaly does not appear irrespective of whether it is incident on a heavy or light target [19, 20, 21, 22, 23, 24, 25, 26, 27, 28]. As for ^{7}Li the conclusions seem to be different since it has been shown that the behaviour does depend on the mass of the target. This point has been discussed with detail elsewhere [19, 21]. As already mentioned, elastic scattering measurements of weakly bound nuclei at energies around the Coulomb barrier are of great importance. For doing theoretical analyse, such as an analysis of the threshold anomaly or breakup threshold anomaly, a good set of data points, that is, complete elastic scattering angular distributions at several energies are always helpful. In this context, it came to our attention that the information existent on elastic scattering angular distributions covering a wide angular range for ^{7}Li on medium targets at energies below the Coulomb barrier is very scarce. Most of the measurements reported correspond to energies near but above the Coulomb barrier; ^{7}Li+^{58}Ni [29], ^{7}Li+^{59}Co [26], ^{7}Li+^{64}Ni [21]. In addition, the measurements reported for energies near but below the Coulomb barrier for ^{7}Li+^{58}Ni are not full angular distributions as reported by Zerva K *et al.* [30], where the elastic backscattering technique is used. With this in mind, recently the Heavy-Ion Group made full experimental elastic scattering measurements for the system ^{7}Li+^{58}Ni at ten different energies around and below the Coulomb barrier (V_b = 14.0 MeV in the laboratory frame of reference). The measurements were made at the Tandem Particle Accelerator Laboratory in the National Institute for Nuclear Research (ININ) in Mexico. In this work, we present preliminary angular distributions of ^{7}Li elastically scattered from ^{58}Ni for five out of the ten energies measured. Preliminary calculations using the optical model potential (OMP) are also reported as well as a comparison with data reported by Pfeiffer K O *et al.* [29] and by Zerva K *et al.* [30].

2. Experimental setup

The experiment took place at the particle accelerator laboratory in the National Institute for Nuclear Research in Mexico. This particular accelerator, used during the experiment, is a Tandem Van de Graaff with a nominal maximum operational voltage of 6 MV and is the only one in Mexico that produces stable beams of heavy ions (from hydrogen to carbon) at energies around the barrier, of great interest depending on the system. The ^{7}Li beam current, produced by the particle accelerator, had intensities from 10 to 20 pnA depending on the energy. Two different targets were used, one of them was a gold foil (^{197}Au) with an approximate thickness of 1.11 mg/cm^2 and the other one was an enriched (99.9%) ^{58}Ni foil with a nominal thickness of 0.259 mg/cm^2. The gold target was used for solid angle determination. The measurements were made in a 70 cm diameter scattering chamber (experimental line of 30°), at ten different energies being 9.0 and 14.22 MeV (laboratory frame of reference) the lowest and highest energy, respectively, with steps of 0.5 MeV between 10 and 13.5 MeV. Inside the chamber (2π geometry), a circular table and a target holder are located, and both of them can be rotated to any desired angle. The detectors can be placed on the circular table at different angular positions. The size of the beam was defined by a set of vertical and horizontal slits placed on the wall of the chamber with an aperture of 3 mm × 3 mm.

The scattered particles were detected with six silicon surface-barrier (SSB) detectors. Four of them were placed at forward angles and two at backward angles with respect to the beam axis as shown in figure 1. In a first experimental setup (configuration) the position of the normal to the target was set at 10° with respect to the ^{7}Li beam axis, and the positions of each of the six detectors were 35° (det. A), 45° (det. B), 55° (det. C), 65° (det. D), 155° (det. E) and 165° (det. F), respectively. After the measurements were made with this first configuration, the circular table and target holder were rotated to a second configuration, leaving the target at 40° and the detectors at 65° (det. A), 75° (det. B), 85° (det. C), 95° (det. D), 120° (det. E) and 135° (det. F) with respect to the beam axis, respectively. This procedure was repeated for all the energies measured. The solid angles subtended by the detectors at the target center were 2.25, 5.69, 2.03, 1.90, 7.26 and 3.3 msr for detectors A, B, C, D, E and F, respectively.

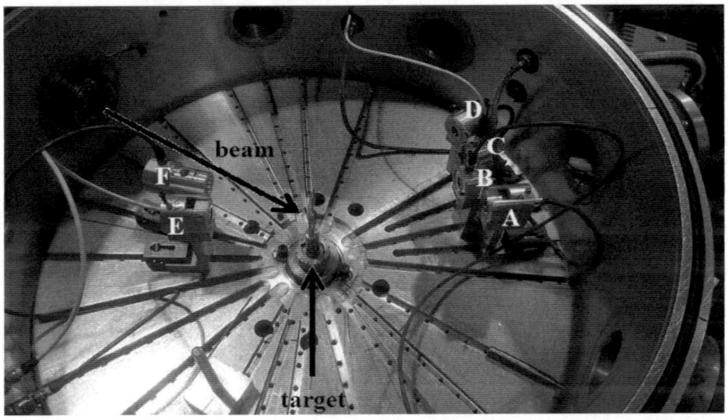

Figure 1. Scattering chamber showing the experimental setup.

3. Results and discussion

The preliminary elastic scattering angular distributions measured at the energies of 12.0, 12.5, 13.0, 13.5 and 14.22 MeV in the laboratory frame of reference are shown in figure 2. The total error reported considers the statistical and systematic errors which in some cases are smaller than the symbols as shown in the figure. These preliminary angular distributions were normalized to Rutherford scattering at the lowest angle measured to allow the determination of deviations from pure Coulomb scattering.

The experimental data were analyzed using the double-folding São Paulo Potential (SPP) [31, 32] for the real part of the nuclear interaction, and a standard Woods-Saxon (WS) shape for the imaginary part. The values of the potential parameters obtained for the imaginary part are indicated in table 1, where W is the depth, R_i is the radius and a_i the diffuseness of the potential. The respective calculations are represented by the curves shown in figure 2. For all five energies it was possible to fit the data using the same value for the parameters of the imaginary part of the potential, as shown in table 1. It can be seen from figure 2 that the calculations obtained in this work describe very well the preliminary experimental data for the ^{7}Li+^{58}Ni system. All χ^2/N values reported here refer to χ^2 per point. The total reaction cross sections (σ_R) for each energy are also given in table 1.

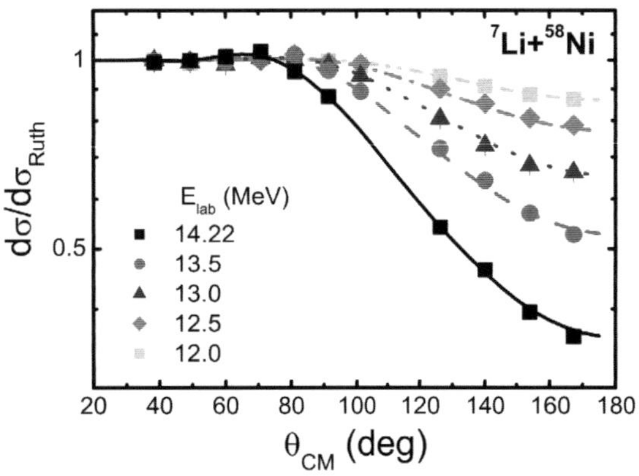

Figure 2. Preliminary elastic scattering angular distributions for ^{7}Li+^{58}Ni at the five energies indicated. The curves correspond to optical model calculations using the São Paulo Potential (SPP) for the real part and a Woods-Saxon shape for the imaginary part (the parameters are shown in table 1).

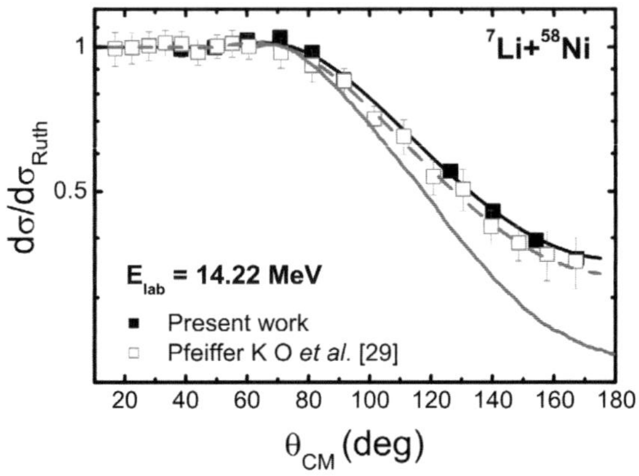

Figure 3. Comparison of the elastic scattering angular distribution for E_{lab} = 14.22 MeV between data reported in the present work and those by Pfeiffer K O *et al.* [29]. The curves (black solid line and red dashed line) correspond to optical model calculations using the SPP for the real part and a Woods-Saxon shape for the imaginary part for each set of experimental points (the parameters are shown in table 1). The red solid line corresponds to optical model calculations reported in Ref. [29], see text for further information.

Table 1. Optical-model potential obtained for ^{7}Li+^{58}Ni from the present data and the calculated total reaction cross sections for each energy. The SPP is used for the real part and a Woods-Saxon shape for the imaginary part.

E_{lab} (MeV)	W (MeV)	R_i (fm)	a_i (fm)	χ^2/N	σ_R (mb)
12.0	60	1.2	0.56	0.18	32 ± 13
12.5	60	1.2	0.56	0.46	58 ± 12
13.0	60	1.2	0.56	0.75	96 ± 12
13.5	60	1.2	0.56	1.06	148 ± 12
14.22	60	1.2	0.56	0.56	239 ± 13
14.22*	60	1.2	0.60	0.08	281 ± 84

* experimental data reported by Pfeiffer K O *et al.* [29]

As part of the analysis presented in this work, a comparison of the elastic scattering angular distribution for E_{lab} = 14.22 MeV between data reported by Pfeiffer K O *et al.* [29] and the present work was made, as shown in figure 3. From this figure, it is possible to see that the preliminary experimental data reported in this work, for this energy, are in good agreement with respect to the data reported back in 1973 by Pfeiffer K O *et al.* [29]. The latter data were also analized using the SPP for the real part and a WS shape for the imaginary part of the potential (red dashed line), the respective parameters are also shown in table 1. It can be seen from the table that the only parameter that had to be modified in order to get the best fit for these experimental data was the diffuseness (from 0.56 to 0.6 fm). The total reaction cross section obtained with the SPP for the real part and the WS for the imaginary part of the potential gives a value of 281 mb while Pfeiffer K O *et al.* [29] reports a value of 420 mb. In figure 3, it is possible to observe that the elastic scattering angular distribution fit (red solid line) reported in Ref. [29] fails to reproduce the experimental data at angles above 80° (center of mass reference frame), which explains the large difference in the above reaction cross section values. The red solid curve of figure 3 was directly extracted from figure 4b of Ref. [29]. In Pfeiffer's work, a normal volume Woods-Saxon shape for the real and imaginary parts of the optical potential was used to analyze the experimental elastic scattering angular distributions (see table 3 of Ref. [29]). The WS parameters reported by Pfeiffer for the energy of 14.22 MeV were $V = 152$ MeV, $r_r = 1.32$ fm and $a_r = 0.79$ fm for the real part, and $W = 3.65$ MeV, $r_i = 2.48$ fm and $a_i = 0.49$ fm for the imaginary part. Throughout Pfeiffer's analysis the value of the radius was calculated as $R = r_0 A_t^{1/3}$, where A_t is the target mass and the value of the reduced radius used was $r_0 = 1.25$ fm. Using these parameter values, we were able to reproduce the red solid line of figure 3.

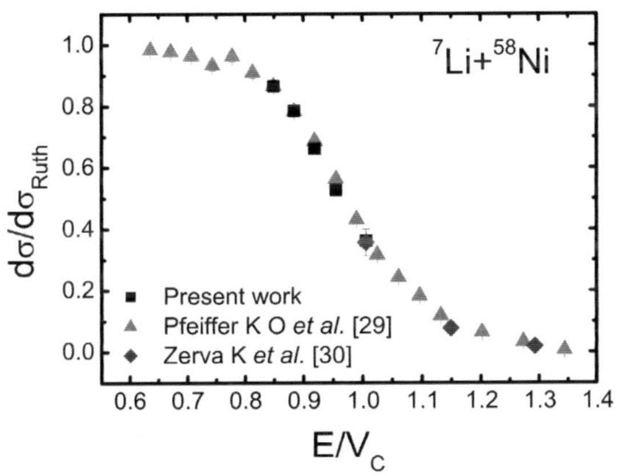

Figure 4. Comparison of the excitation functions of average cross sections between 160° and 170° for the ^{7}Li+^{58}Ni system. The experimental data points reported in the present work and those reported by Ref. [29] are measurements taken from full angular distributions.

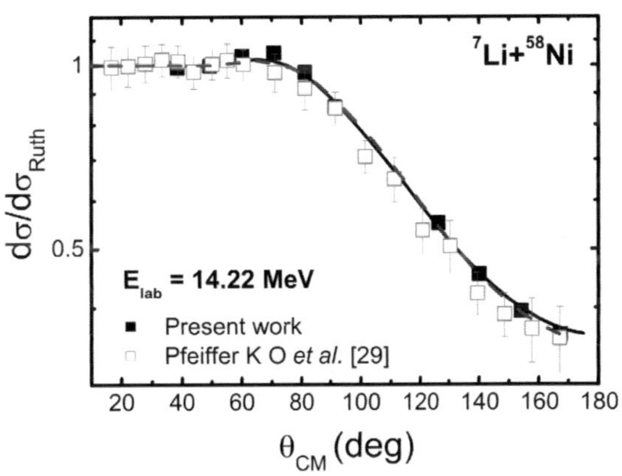

Figure 5. Elastic scattering angular distribution for E$_{lab}$ = 14.22 MeV reported in this work and by Pfeiffer K O *et al.* [29]. Calculations using the São Paulo Potential (SPP) for the real part and a Woods-Saxon shape for the imaginary part of the nuclear potential (black solid line) and data extracted from Ref. [30] (blue dashed line), see text for further information.

On the other hand, Zerva K *et al.* [30] performed precision elastic backscattering measurements for the ^{7}Li+^{58}Ni system at energies from 9.0 MeV to 19.0 MeV in the laboratory frame of reference. In that work, the excitation function of average cross-sections between 160° and 170° for the ^{7}Li+^{58}Ni system was reported. Since in Pfeiffer K O *et al.* [29] as well as in the present work, the full elastic scattering angular distributions for ^{7}Li+^{58}Ni are reported, it was possible to extract the excitation function and compare with Zerva's experimental data. This comparison is shown in figure 4. For this comparison, a value of $V_C = 14.14$ MeV (laboratory frame of reference) for the Coulomb barrier of the ^{7}Li+^{58}Ni system was used, as indicated in Ref. [30]. When comparing the excitation function from the present work with those from Pfeiffer K O *et al.* [29] and Zerva K *et al.* [30] it is possible to see that the data points are in good agreement in the small angular range considered (160° – 170°).

In addition, Zerva K *et al.* compared the elastic scattering angular distributions reported in 1973 by Pfeiffer K O *et al.* with optical model predictions using the potentials obtained via the backscattering barrier distribution technique (see Ref. [30] for details). The mentioned predictions (blue dashed line) for $E_{lab} = 14.22$ MeV reported by Zerva as well as the experimental data reported by Pfeiffer and the preliminary experimental data reported in this work are shown in figure 5. The prediction found by Zerva fails to reproduce Pfeiffer's experimental data as shown in figure 5. This in fact is indicated in Zerva's work, since it is stated that despite several trial attempts it was impossible to find a potential which described the experimental data reported by Pfeiffer *et al.* and that this was a possible indication that new angular distribution measurements may be necessary. Following Zerva's statement and as part of the motivation for this experiment, it can be seen from the same figure that the preliminary experimental data for $E_{lab} = 14.22$ MeV reported in this work are well reproduced by the prediction reported in Zerva's work. At the same time, the optical model predictions obtained in the present work are also in agreement with Zerva's prediction. This can be seen as an indication that the new experimental data, for the system ^{7}Li+^{58}Ni, is a reliable set of data points which may be used for analysis of the threshold anomaly or breakup threshold anomaly.

4. Conclusions

Recently, measurements of ^{7}Li elastically scattered from ^{58}Ni at energies near and below the Coulomb barrier were made. In the present work, preliminary elastic scattering angular distributions for five out of ten energies were reported. By using the double-folding São Paulo potential for the real part of the nuclear interaction and a Woods-Saxon shape for the imaginary part, the five preliminary angular distributions were well described. The comparison between full and partial (160° – 170°) experimental angular distributions for $E_{lab} = 14.22$ MeV reported back in 1973 [29] and in 2012 [30], respectively, are in good agreement with the full preliminary data reported in the present work. Further analysis is in progress in order to further understand this particular system.

Acknowledgments

The authors would like to acknowledge the staff of the Tandem Van de Graaff particle accelerator at ININ for all their support and time dedication during this experiment. This work was partially supported by CONACYT.

References

[1] Canto L F, Gomes P R S, Donangelo R and Hussein M S 2006 *Phys. Rep.* **424** 1
[2] Keeley N, Raabe R, Alamanos N and Sida J L 2007 *Part. Nucl. Phys.* **59** 579
[3] Keeley N, Alamanos N, Kemper K and Rusek K 2009 *Part. Nucl. Phys.* **63** 396
[4] Aguilera E F *et al.* 2000 *Phys. Rev. Lett.* **84** 5058
[5] Aguilera E F *et al.* 2001 *Phys. Rev.* C **63** 061603(R)
[6] Kolata J J *et al.* 1998 *Phys. Rev. Lett.* **81** 4580
[7] De Young P A *et al.* 2005 *Phys. Rev.* C **71** 051601(R)
[8] Guimarães V *et al.* 2000 *Phys. Rev. Lett.* **84** 1862
[9] Aguilera E F *et al.* 2009 *Phys. Rev.* C **79** 021601(R)
[10] Aguilera E F *et al.* 2011 *Phys. Rev. Lett.* **107** 092701
[11] Pakou A *et al.* 2013 *Phys. Rev.* C **87** 014619

[12] Fernández-García J P , Rodríguez-Gallardo M, Alvarez M A G and Moro A M 2010 *Nucl. Phys.* A **840** 19–38
[13] Sánchez-Benítez A M *et al.* 2008 *Nucl. Phys.* A **803** 30–45
[14] Milin M *et al.* 2004 *Nucl. Phys.* A **730** 285–298
[15] Kakuee O R *et al.* 2003 *Nucl. Phys.* A **728** 339–349
[16] Benjamim E A *et al.* 2007 *Phys. Lett.* B **647** 30–35
[17] Satchler G R 1991 *Phys. Rep.* **199** 147
[18] Gomes P R S and Lubian J 2014 *J. Phys.: Conf. Series* **533** 012029
[19] Di Pietro A 2010 *J. Phys.: Conf. Series* **205** 012042
[20] Hussein M S *et al.* 2006 *Phys. Rev.* C **73** 044610
[21] Shaikh Md. Moin, Das Mili, Roy Subinit, Sinha M, Pradhan M K, Basu P, Datta U, Ramachandran K, Shrivastava A 2016 *Nucl. Phys.* A **953** 80–94
[22] Keeley N, Bennett S J, Clarke N M, Fulton B R, Tungate G, Drumm P V, Nagarajan M A and Lilley J S, 1994 *Nucl. Phys.* A **571** 326
[23] Maciel A M M *et al.* 1999 *Phys. Rev.* C **59** 2103
[24] Pakou A *et al.* 2004 *Phys. Rev.* C **69** 054602
[25] Figueira J M *et al.* 2007 *Phys. Rev.* C **75** 017602
[26] Souza F A, Leal L A S, Carlin N, Munhoz M G, Liguori Neto R, de Moura M M, Suaide A A P, Szanto E M, Szanto de Toledo A and Takahashi J 2007 *Phys. Rev.* C **75** 044601; 2007 *Phys. Rev.* C **76** 029901(E)
[27] Biswas M, Roy Subinit, Sinha M, Pradhan M K, Mukherjee A, Basu P, Majumdar H, Ramachandran K and Shrivastava A 2008 *Nucl. Phys.* A **802** 67–81
[28] Deshmukh N N *et al.* 2011 *Phys. Rev.* C **83** 024607
[29] Pfeiffer K O, Speth E and Bethge K 1973 *Nucl. Phys.* A **206** 545–557
[30] Zerva K *et al.* 2012 *Eur. Phys. J.* A **48** 102
[31] Chamon L C, Pereira D, Hussein M S, Cândido Ribeiro M A and Galetti D (1997) *Phys. Rev. Lett.* **79** 5218
[32] Chamon L C, Carlson B V, Gasques L R, Pereira D, De Conti C, Alvarez M A G, Hussein M S, Cândido Ribeiro M A, Rossi Jr. E S and Silva C P 2002 *Phys. Rev.* C **66** 014610
[33] Thompson I J 1988 *Comput. Phys. Rep.* **7** 167–212

Overview of the data analysis and new micro-pattern gas detector development for the Active Target Time Projection Chamber (AT-TPC) project.

Yassid Ayyad[1], Wolfgang Mittig, Daniel Bazin and Marco Cortesi

National Superconducting Cyclotron Laboratory, Michigan State University, East Lansing, MI 48824, USA

E-mail: ayyad@lbl.gov

Abstract. The Active Target Time Projection Chamber (AT-TPC) project at the NSCL (National Superconducting Cyclotron Laboratory, Michigan State University) is a novel active target detector tailored for low-energy nuclear reactions in inverse kinematics with radioactive ion beams. The AT-TPC allows for a full three dimensional reconstruction of the reaction and provides high luminosity without degradation of resolution by the thickness of the target. Since all the particles (and also the reaction vertex) are tracked inside the detector, the AT-TPC has full 4π efficiency. The AT-TPC can operate under a magnetic field (2 T) that improves the identification of the particles and the energy resolution through the measurement of the magnetic rigidity. Another important characteristic of the AT-TPC is the high-gain operation achieved by the hybrid thick Gas Electron Multipliers (THGEM)-Micromegas pad plane, that allow operation also in pure elemental gas. These two features make the AT-TPC a unique high resolution spectrometer with full acceptance for nuclear physics reactions. This work presents an overview of the project, focused on the data analysis and the development of new micro-pattern gas detectors.

1. Introduction

Time projection chambers operated in active target mode have recently gained popularity with the advent of new facilities dedicated to produce very exotic radioactive beams [1, 2]. In active targets the filling gas is simultaneously the tracking medium and the target. The latter is commonly a light gas such as hydrogen, deuterium, helium or argon. Inside the detection medium, particles leave an ionization trace that can be discretized by measuring two projected coordinates in a highly segmented pad plane and by the drift time of the ionized electrons. The characteristic energy loss pattern and range of the particles is also recorded. The addition of a solenoidal magnetic field adds another variable in the reconstruction of the kinematics of the reaction by measuring the magnetic rigidity of the particle. Under such conditions active targets become a powerful solenoidal spectrometer with full acceptance, large dynamic range and detection efficiency. In addition, the typical excitation energy degradation caused by the reaction target thickness is dramatically reduced. Target thickness can be increased by an order of magnitude with respect to the most common setups with solid targets.

[1] Present address: Nuclear Science Division, Lawrence Berkeley National Laboratory, Berkeley, CA 94720, USA

The AT-TPC, developed at the NSCL [3, 4], is currently one of the thickest reaction targets operating under a magnetic field of 2 T. With a luminosity of around 100 times larger than conventional setups and full geometrical acceptance (depending on the reaction of interest) the AT-TPC allows experiments with beam intensities down to 100-1000 Hz. The read-out of the detector, composed of a hybrid micromegas [5] and thick GEMs [6, 7, 8], enables high-gain operation in pure elemental gas without addition of a quencher that introduces background reactions. The AT-TPC allows the reconstruction of the reaction in three dimensions with high resolution on an event-by-event basis. The complexity underlies on how the track coordinates are transformed into variables of kinematic interest (i.e. angle, energy and time), that usually are inferred directly. Such a complex system demands a sophisticated data analysis framework with the capability to classify the events of interest and reconstruct the kinematics of the particles with high resolution. The most challenging aspects are the large energy loss dynamic range of the particles and the fact that particles that are stopped in the gas will describe highly non-linear trajectories (helices) characterized by a strong change on the radius of curvature as it slows down. We have developed a novel algorithm that classifies the tracks of the particles and permits the inference of the kinematic variables used for further post-processing. In this work, we present the general aspects of the algorithm as well as a brief discussion on new thick GEM technologies developed for the AT-TPC. The detailed description of the AT-TPC detector as well as its performance can be found in Ref. [3].

2. Classification of spiral particle tracks

We conducted a commissioning experiment with a re-accelerated radioactive beam of ^{46}Ar in ReA3 separator of the NSCL. The AT-TPC was filled with isobutane gas (C_4H_{10}) at 18 torr to perform the resonant elastic scattering of ^{46}Ar on protons at 4.6 MeV/u. In the left panel of Fig. 2 we show the three dimensional hit pattern of a proton track in the gas medium within a magnetic field of 2 T. Each point of this hit pattern has been reconstructed by projecting the electron drift signals in the pad plane (composed of 10240 triangular pads), obtaining three cartesian coordinates and the peak amplitude that represents the energy loss per unit volume. As shown in Fig. 2, the tracks are characterized by a helical trajectory with a radius of curvature decreasing due to the energy loss of the particle of the medium. This particular track geometry cannot be described by any analytical parameterization that represents helical trajectories in the space. Thus, a multi-step reconstruction algorithm based on random sampling was implemented in several steps. The whole procedure involves the classification of the tracks and noise reduction, and the fit of each spiral track using a Monte Carlo algorithm. The latter is further explained in a forthcoming publication [3].

For the classification of the spiral tracks we used a global pattern recognition algorithm based on the Hough transformation [9]. The idea of this algorithm is to detect shapes in images by transforming the image data set given in the coordinate space into a parameter space (Hough space) described by the parameters of the shape we want to find. We chose the parameterization of a circle based on a straight line between two neighboring points of the hit pattern and a perpendicular line to it. Such a line intersects the center of the circle if the pair of points belongs to it. This reduces the dimensionality of the problem with only two parameters to consider: the distance of the circle center with respect to the origin of the coordinate system XY and its angle with the positive x-axis. Although the projection of the spiral track into the pad plane does not have the shape of a unique circle, one can see that the total shape is composed by several superimposed circles of different radius of curvature.

By using this parameterization each one of the points of the hit pattern is transformed into a

IOP Conf. Series: Journal of Physics: Conf. Series **876** (2017) 012003 doi:10.1088/1742-6596/876/1/012003

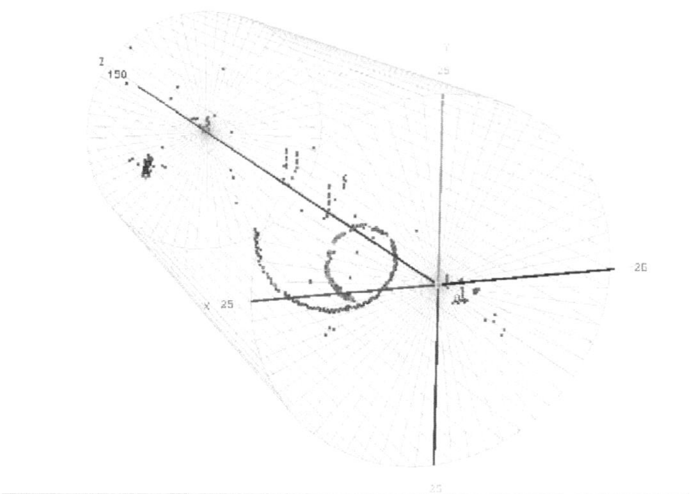

Figure 1. Three-dimensional image of a scattered proton. The coordinates of each point are inferred from the position of the particle in the pad plane and the drift time converted into length. The size of each point of the hit pattern is proportional to the energy loss.

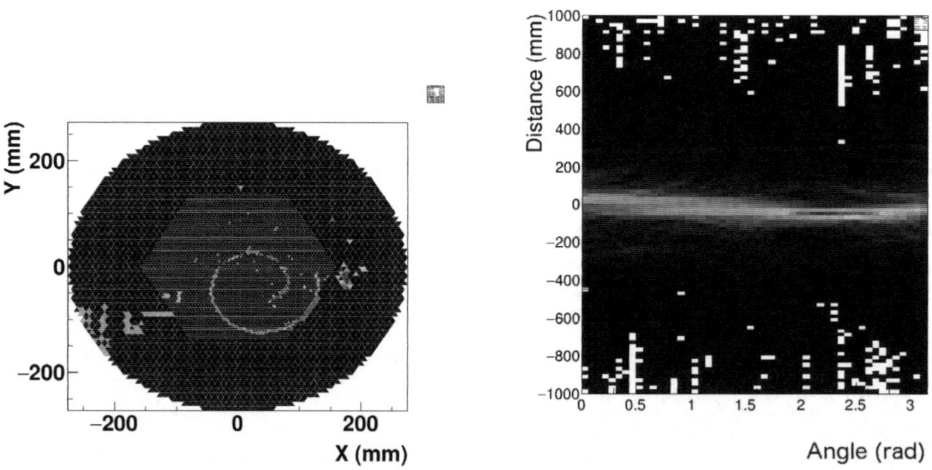

Figure 2. (Left panel) Projection of the track in the XY pad plane. The local energy loss per time is recorded in each activated pad. (Right panel) Distance-angle parameter space. The parameters of the trajectory are determined from the bin with maximum number of counts.

binned two-dimensional function in the distance-angle parameter space (see right panel of Fig. 2). The bin with maximum number of counts gives the parameters of the circle we want to find. Misidentified tracks can be rejected by evaluating the maximum/minimum bin ratio. Once the parameters are determined, we can transform back the point of interest into the coordinate space to infer the radius of curvature and the magnetic rigidity, the polar (scattering) and azimuthal angles, and the vertex of the reaction. These starting parameters allow us to perform a Monte

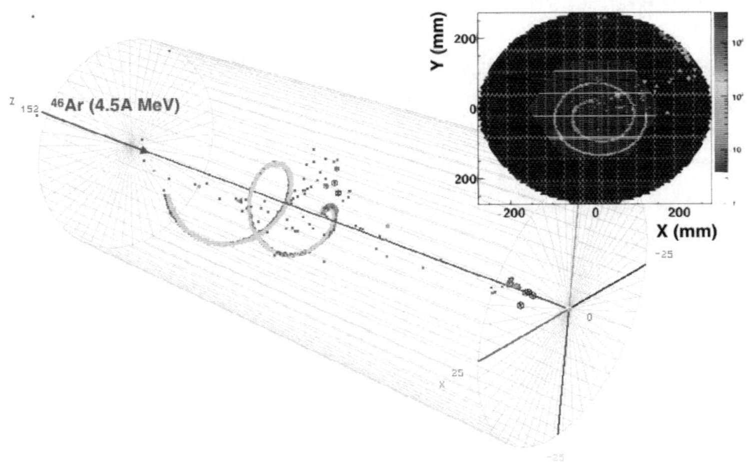

Figure 3. Reconstructed proton track. Red boxes represent the hit pattern while the green dots refer to the reconstructed track. The inset shows the projection in the XY pad plane.

Carlo fit to determine the energy and the scattering angle of the particles with higher precision. The fitting procedure is described elsewhere [3]. The fit algorithm generates a large number of possible tracks with a simulation of the signals induced in the pads. The simulated tracks are compared to the energy loss and hit patterns of the real track. The χ^2 between the simulated and experimental tracks is calculated and minimized. The energy loss of the particles was computed using SRIM code [10]. The reconstructed track is shown in Fig. 3 as green dots. Together with the pattern recognition task, we apply a noise reduction filter based on the statistical analysis of each point. We used the Statistical Outlier Removal filter from the Point Cloud Library [11] prior to the Hough space calculation. The AT-TPC collaboration is developing a modular data analysis framework that includes several tools to unpack, sort and analyze the data taken with the detector [12].

3. Development of new thick Gas Electron Multiplier (GEM) based on ceramic substrates.

As mentioned in the previous sections, one of the important features of the AT-TPC is the use of advanced GEM-like multiplier devices that allow us to operate the detector with pure noble gases. A novel charge-avalanche concept was recently introduced, multi-layer THGEM (M-THGEM) [7]. The M-THGEM is composed of a stack of two or three conventional THGEM, one on top of the other. The diameter and pitch of the holes are 0.5 mm and 1 mm, respectively. These detectors showed an excellent performance when operated in low-pressure pure noble gas, achieving comparable gains than the ones obtained with mixtures such as He+CO_2. The main reason is the strong confinement of the avalanche inside the M-THGEM hole, which leads to a effective reduction of the photo-mediated secondary effects.

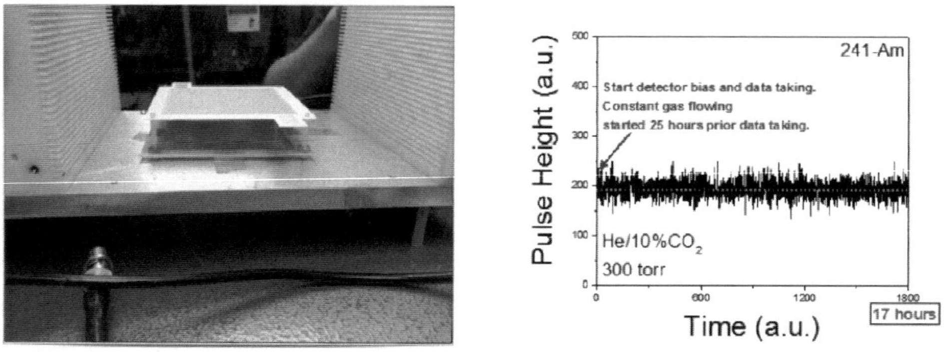

Figure 4. (Left panel) Ceramic thick GEM mounted on the test chamber. (Right panel) Pulse height as a function of the time.

One of the most common problems present in GEM-like multipliers, and in gaseous detectors in general, are long-term variations caused by charge collected on the dieletric surface during the avalanche process (charging up effects). These effects are a complex phenomenon, which involve several factors, including certain impurities in the substrates that affect the surface resistivity of the insulator or very different substrate materials combined together. The FR-4/G10 material used for conventional THGEMs is made of glass fiber and epoxy plus many other chemical compounds and impurities (including water) that increase charging up effects. On the other hand, other materials such as ceramic have less impurities and absorb lower amounts of water. In addition, the surface of the FR-4/G10 is quite irregular. Ceramic substrates (especially with laser production and suitable cleaning procedure for removing impurities and humidity) lead to a lower amount of water and a more regular surface, resulting in a lower (not fully suppressed) charge building up effects. Therefore, we developed thick GEMs based on ceramic substrates. A three-layer 10×10 cm^2 thick GEM prototype was mounted together with a single readout pad plane, to pick-up the signal of the avalanche, inside a vacuum chamber. Our preliminary measurements using a mixture of He+CO_2 gas at 300 torr, show a stable operation during a period of 17 h when irradiated with a ^{241}Am source (see Fig. 4) at relative low rate (a few hundred Hz). In addition, we measured an energy resolution from the collected charge of around 5% (FWHM). These results will be published in a separated paper.

4. Conclusions

We have developed a novel data analysis framework for the Active Target Time Projection Chamber (AT-TPC) project at the NSCL that includes pattern recognition algorithms coupled to a Monte Carlo fit. The pattern recognition algorithm, based on the Hough transformation, is used to classify the tracks and extract the kinematical parameters from them. Using these parameters as initial starting point, we perform a fit using a Monte Carlo optimization algorithm. We also present a new multi-layer thick GEM device based on ceramic substrate that shows a better stability against charging up effects and high-gain operation in low-pressure pure elemental gas.

References

[1] S. Beceiro-Novo *et al.*, "Active targets for the study of nuclei far from stability," *Progress in Particle and Nuclear Physics*, vol. 84, pp. 124 – 165, 2015.

IOP Conf. Series: Journal of Physics: Conf. Series **876** (2017) 012003 doi:10.1088/1742-6596/876/1/012003

[2] W. Mittig *et al.*, "Active target detectors for studies with exotic beams: Present and next future," *Nuclear Instruments and Methods in Physics Research Section A: Accelerators, Spectrometers, Detectors and Associated Equipment*, vol. 784, pp. 494 – 498, 2015. Symposium on Radiation Measurements and Applications 2014 (SORMA XV).

[3] J. Bradt *et al.*, "The active-target time projection chamber," *Nuclear Instruments and Methods in Physics Research Section A: Accelerators, Spectrometers, Detectors and Associated Equipment*, 2017. submitted.

[4] D. Suzuki *et al.*, "Prototype at-tpc: Toward a new generation active target time projection chamber for radioactive beam experiments," *Nuclear Instruments and Methods in Physics Research Section A: Accelerators, Spectrometers, Detectors and Associated Equipment*, vol. 691, pp. 39 – 54, 2012.

[5] Y. Giomataris *et al.*, "Micromegas: a high-granularity position-sensitive gaseous detector for high particle-flux environments," *Nuclear Instruments and Methods in Physics Research Section A: Accelerators, Spectrometers, Detectors and Associated Equipment*, vol. 376, no. 1, pp. 29 – 35, 1996.

[6] M. Cortesi *et al.*, "Studies of thgem-based detector at low-pressure hydrogen/deuterium, for at-tpc applications," *Journal of Instrumentation*, vol. 10, no. 09, p. P09020, 2015.

[7] M. Cortesi *et al.*, "Multi-layer thick gas electron multiplier (m-thgem): A new mpgd structure for high-gain operation at low-pressure," *Review of Scientific Instruments*, vol. 88, no. 1, p. 013303, 2017.

[8] A. Breskin *et al.*, "A concise review on {THGEM} detectors," *Nuclear Instruments and Methods in Physics Research Section A: Accelerators, Spectrometers, Detectors and Associated Equipment*, vol. 598, no. 1, pp. 107 – 111, 2009.

[9] J. Illingworth and J. Kittler, "The adaptive hough transformation," *IEEE Transactions on Pattern Analysis and Machine Intelligence*, vol. 9, pp. 690 – 698, 1987.

[10] J. F. Ziegler *et al.*, "{SRIM} - the stopping and range of ions in matter (2010)," *Nuclear Instruments and Methods in Physics Research Section B: Beam Interactions with Materials and Atoms*, vol. 268, no. 11 - 12, pp. 1818 – 1823, 2010. 19th International Conference on Ion Beam Analysis.

[11] "Point cloud library." https://github.com/PointCloudLibrary/pcl/releases/tag/pcl-1.8.0.

[12] Y. Ayyad *et al.*, "In preparation." https://github.com/ATTPC/ATTPCROOTv2.

IOP Conf. Series: Journal of Physics: Conf. Series **876** (2017) 012004 doi:10.1088/1742-6596/876/1/012004

Pentaquark states with hidden charm

Roelof Bijker

Instituto de Ciencias Nucleares, Universidad Nacional Autónoma de México, A.P. 70-543, 04510 México, D.F., México

E-mail: bijker@nucleares.unam.mx

Abstract. I develop an extension of the usual three-flavor quark model to four flavors (u, d, s and c), and discuss the classification of pentaquark states with hidden charm. This work is motivated by the recent observation of such states by the LHCb Collatoration at CERN.

1. Introduction

The recent observation of candidates for pentaquark states by the LHCb Collaboration [1, 2, 3] has generated an enormous renewed interest in these multiquark states with almost weekly publications in the arXive and the scientific literature. The first evidence was found in the $\Lambda_b^0 \to J/\psi p K^-$ decay [1]. An analysis of the J/ψ invariant mass spectrum showed the need to introduce two hidden charm pentaquark states, $P_c^+(4380)$ and $P_c^+(4450)$, with opposite parity. The evidence for the existence of these pentaquark states was confirmed later in a model-independent analysis [2], as well as in the $\Lambda_b^0 \to J/\psi p \pi^-$ decay [3]. So far, the angular momentum and parity could not be determined uniquely. The most likely values are $J^P = 3/2^-$ for $P_c^+(4380)$, and $5/2^+$ for $P_c^+(4450)$, although other combinations, like ($3/2^+$, $5/2^-$) or ($5/2^+$, $3/2^-$), are not excluded. There are proposals to study the $P_c^+(4380)$ and $P_c^+(4450)$ states in photoproduction of J/ψ on the proton to verify their existence and to determine their spin and parity [4]-[9].

Theoretically, several narrow N^* and Λ^* resonances with hidden charm were predicted above 4 GeV in a coupled-channel unitary approach [10, 11] several years before the LHCb data, as well as [12, 13]. After the discovery of the P_c^+ pentaquark states many different explanations have been explored, ranging from weakly bound molecular states of charmed baryons and mesons [13]-[19], and interpretations in terms of multiquark degrees of freedom [20]-[26], to kinematic effects [27, 28, 29]. More information on the experimental and theoretical aspects of pentaquark states, as well as a more complete list of references, can be found in the reviews [30, 31, 32].

It is the aim of the present contribution to present a classification of all possible $uudc\bar{c}$ hidden-charm pentaquark states in an extension of the three- to the four-flavor $SU(4)$ quark model, and to calculate the corresponding magnetic moments of these configurations. This work is an extension of earlier studies of $uudd\bar{s}$ pentaquarks [33, 34].

2. Four-flavor SU(4) quark model

In this section, I review the four-flavor $SU(4)$ quark model which is based on the flavors up, down, strange and charm [35]-[38]. Even though the $SU(4)$ flavor symmetry is expected to be

IOP Conf. Series: Journal of Physics: Conf. Series **876** (2017) 012004 doi:10.1088/1742-6596/876/1/012004

much more broken than the three-flavor $SU(3)$ flavor symmetry due to the large mass of the charm quark, it nevertheless provides a useful classification scheme for baryons and mesons.

The spin-flavor states of the $udsc$ quark model can be decomposed into their flavor and spin parts as

$$\left| \begin{array}{ccccc} SU_{\rm sf}(8) & \supset & SU_{\rm f}(4) & \otimes & SU_{\rm s}(2) \\ {[f]} & , & [g] & , & S \end{array} \right\rangle , \tag{1}$$

followed by a reduction to the three-flavor uds states

$$\left| \begin{array}{ccccc} SU_{\rm f}(4) & \supset & SU_{\rm f}(3) & \otimes & U_{\rm Z}(1) \\ {[g]} & , & [h] & , & Z \end{array} \right\rangle , \tag{2}$$

and finally to their isospin and hypercharge contents

$$\left| \begin{array}{ccccc} SU_{\rm f}(3) & \supset & SU_{\rm I}(2) & \otimes & U_{\rm Y}(1) \\ {[h]} & , & I & , & Y \end{array} \right\rangle . \tag{3}$$

The quarks transform as the fundamental representation [1] under $SU(n)$, where $n = 2$ for the spin, $n = 3$ for the color, $n = 4$ for the flavor, and $n = 8$ for the combined spin-flavor degrees of freedom, whereas the antiquarks transform as the conjugate representation $[1^{n-1}]$ under $SU(n)$. The hypercharges Y and Z are related to the baryon number B, the strangeness $\mathcal{S}$ and charm C by [32]

$$\begin{aligned} Y &= B + \mathcal{S} - \frac{C}{3} , \\ Z &= \frac{3}{4}B - C , \end{aligned} \tag{4}$$

and the electric charge Q is given by the generalized Gell-Mann-Nishijima relation

$$Q = I_3 + \frac{B + \mathcal{S} + C}{2} . \tag{5}$$

3. $qqqq\bar{q}$ Pentaquark states

The pentaquark wave functions depend both on the spatial degrees of freedom and the internal degrees of freedom of color, flavor and spin

$$\psi = \psi^{\rm o}\phi^{\rm f}\chi^{\rm s}\psi^{\rm c} . \tag{6}$$

In order to classify the corresponding states, we shall make use as much as possible of symmetry principles. In the construction of the classification scheme we are guided by two conditions: (i) the pentaquark wave function should be antisymmetric under any permutation of the four quarks, and (ii) as all physical states, it should be a color singlet. In addition, we restrict ourselves to flavor multiplets containing $uudc\bar{c}$ configurations without orbital excitations and with spin and parity $J^P = S^P = 3/2^-$.

The permutation symmetry among the four quarks is characterized by the S_4 Young tableaux [4], [31], [22], [211] and [1111] or, equivalently, by the irreducible representations of the tetrahedral group $\mathcal{T}_d$ (which is isomorphic to the permutation group $\mathcal{T}_d \sim S_4$) as A_1, F_2, E, F_1 and A_2, respectively.

The total pentaquark wave function has to be a [222] color-singlet state. Since the color wave function of the antiquark is a [11] anti-triplet, the color wave function of the four-quark

Table 1. Spin-flavor decomposition of q^4 states. The subindices denote the dimension of the multiplet.

$SU_{\rm sf}(8)$	$\supset$	$SU_{\rm f}(4)$	$\otimes$	$SU_{\rm s}(2)$	
$[f]$	$\supset$	$[g]$	$\otimes$	$[g']$	$\psi^{\rm sf}_{F_2}$
$[31]_{630}$		$[4]_{35}$	$\otimes$	$[31]_3$	$[\phi_{A_1} \times \chi_{F_2}]_{F_2}$
		$[31]_{45}$	$\otimes$	$[4]_5$	$[\phi_{F_2} \times \chi_{A_1}]_{F_2}$
				$[31]_3$	$[\phi_{F_2} \times \chi_{F_2}]_{F_2}$
				$[22]_1$	
		$[22]_{20}$	$\otimes$	$[31]_3$	$[\phi_E \times \chi_{F_2}]_{F_2}$
		$[211]_{15}$	$\otimes$	$[31]_3$	$[\phi_{F_1} \times \chi_{F_2}]_{F_2}$
				$[22]_1$	

Table 2. $SU_{\rm f}(4) \supset SU_{\rm f}(3) \otimes U_{\rm Z}(1)$ flavor classification of four-quark states (here $q = u,\ d,\ s$).

$SU_{\rm f}(4)$	$\supset$	$SU_{\rm f}(3)$								
$[g]$	$\supset$	$[h]$								
$[4]_{35}$	$\supset$	$[4]_{15}$	$\oplus$	$[3]_{10}$	$\oplus$	$[2]_6$	$\oplus$	$[1]_3$	$\oplus$	$[0]_1$
$[31]_{45}$	$\supset$	$[31]_{15}$	$\oplus$	$[3]_{10} \oplus [21]_8$	$\oplus$	$[2]_6 \oplus [11]_3$	$\oplus$	$[1]_3$		
$[22]_{20}$	$\supset$	$[22]_6$	$\oplus$	$[21]_8$	$\oplus$	$[2]_6$				
$[211]_{15}$	$\supset$	$[211]_3$	$\oplus$	$[21]_8 \oplus [111]_1$	$\oplus$	$[11]_3$				
		$qqqq$ $Z = 1$		$qqqc$ $Z = 0$		$qqcc$ $Z = -1$		$qccc$ $Z = -2$		$cccc$ $Z = -3$

configuration has to be a $[211]$ triplet with F_1 symmetry under $\mathcal{T}_d$. The total q^4 wave function is antisymmetric (A_2), hence the orbital-spin-flavor part is a $[31]$ state with F_2 symmetry

$$\psi \;=\; \left[\psi^{\rm c}_{F_1} \times \psi^{\rm osf}_{F_2}\right]_{A_2} , \tag{7}$$

where the subindices refer to the symmetry properties of the four-quark system under permutation. Moreover, it is assumed that the orbital part of the pentaquark wave function corresponds to the ground state (without orbital excitations) with A_1 symmetry. As a consequence the spin-flavor part of the four-quark wave function has F_2 symmetry

$$\psi^{\rm osf}_{F_2} \;=\; \left[\psi^{\rm o}_{A_1} \times \psi^{\rm sf}_{F_2}\right]_{F_2} . \tag{8}$$

The flavor and spin content of the only allowed spin-flavor configuration, $[31]$ with F_2 symmetry, is given in Table 1. There are four different flavor multiplets $[g] = [4]$, $[31]$, $[22]$ and $[211]$. For pentaquark states with $J^P = 3/2^-$ without orbital excitations the allowed values of the spin of the four-quark system are $S = (g'_1 - g'_2)/2 = 1$ and 2. In the last column, we show the allowed spin-flavor configurations, where ϕ and χ denote the flavor and spin wave functions, respectively.

Since the current interest is in four-quark states with one charm quark, it is convenient to make a decomposition of four into three flavors according to $SU_{\rm f}(4) \supset SU_{\rm f}(3) \otimes U_{\rm Z}(1)$. Table 2 shows that the states with one charm quark have $Z = 0$ and belong to either an $SU(3)$ decuplet ($[h] = [3]$), octet ($[21]$) or singlet ($[111]$). Since the flavor singlet corresponds to a *udsc*

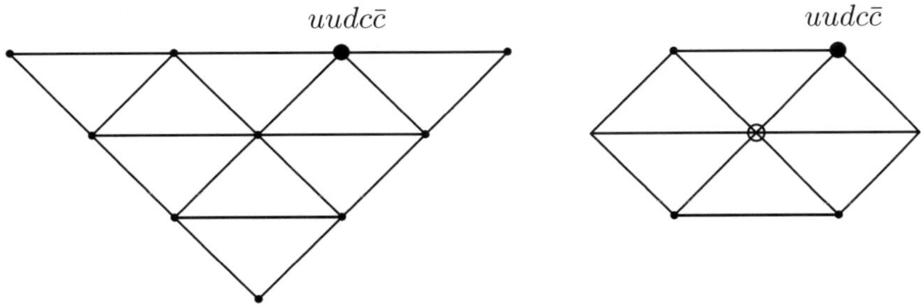

Figure 1. Pentaquark decuplet and octet

configuration, the only $SU(3)$ flavor multiplets that contain a $uudc$ state are the decuplet and the octet.

Finally, the $uudc\bar{c}$ pentaquark states belong to $SU(3)$ multiplets that are obtained by combining the four-quark [3] decuplet and [21] octet with the [111] singlet for the antiquark with charm $\bar{c}$

$$\begin{aligned} [3] \otimes [111] &= [411] \, , \\ [21] \otimes [111] &= [321] \, . \end{aligned} \tag{9}$$

In summary, the $uudc\bar{c}$ pentaquark states belong to a $SU(3)$ flavor decuplet with $[h] = [411] \equiv [3]$ or an octet $[321] \equiv [21]$ (see Fig. 1). The corresponding flavor wave functions can be expressed in terms of the quantum numbers

$$\begin{aligned} |\phi_{A_1}\rangle &= |[g], [h]_{qqqq}, [h]_{\bar{q}}; [h], I, I_3, Y, Z\rangle \, , \\ &= \left|[4], [3], [111]; [411], \frac{3}{2}, \frac{1}{2}, 1, \frac{3}{4}\right\rangle \, , \\ |\phi_{F_2}\rangle &= \left|[31], [3], [111]; [411], \frac{3}{2}, \frac{1}{2}, 1, \frac{3}{4}\right\rangle \, , \end{aligned} \tag{10}$$

for the decuplet, and

$$\begin{aligned} |\phi_{F_2}\rangle &= \left|[31], [21], [111]; [321], \frac{1}{2}, \frac{1}{2}, 1, \frac{3}{4}\right\rangle \, , \\ |\phi_{E}\rangle &= \left|[22], [21], [111]; [321], \frac{1}{2}, \frac{1}{2}, 1, \frac{3}{4}\right\rangle \, , \\ |\phi_{F_1}\rangle &= \left|[211], [21], [111]; [321], \frac{1}{2}, \frac{1}{2}, 1, \frac{3}{4}\right\rangle \, , \end{aligned} \tag{11}$$

for the octet.

4. Magnetic moments

In this section, we study the magnetic moments of the $uudc\bar{c}$ pentaquark states with spin and parity $J^P = 3/2^-$ discussed in the previous section. The magnetic moment is a crucial ingredient in calculation of the photoproduction cross sections of pentaquarks. In Ref. [22], the magnetic moments were calculated for a molecular model, a diquark-triquark model and a

Table 3. Magnetic moments of $uudc\bar{c}$ pentaquark states in μ_N.

	State	Magnetic moment	
Decuplet	$[\phi_{A_1} \times \chi_{F_2}]_{F_2}$	$\frac{1}{2}(2\mu_u + \mu_d + \mu_c) + \mu_{\bar{c}}$	1.164
	$[\phi_{F_2} \times \chi_{A_1}]_{F_2}$	$\frac{9}{10}(2\mu_u + \mu_d + \mu_c) - \frac{3}{5}\mu_{\bar{c}}$	3.065
	$[\phi_{F_2} \times \chi_{F_2}]_{F_2}$	$\frac{2}{3}(2\mu_u + \mu_d) + \mu_{\bar{c}}$	1.417
Octet	$[\phi_{F_2} \times \chi_{A_1}]_{F_2}$	$\frac{9}{10}(2\mu_u + \mu_d + \mu_c) - \frac{3}{5}\mu_{\bar{c}}$	3.065
	$[\phi_{F_2} \times \chi_{F_2}]_{F_2}$	$\frac{1}{12}(14\mu_u + \mu_d + 9\mu_c) + \mu_{\bar{c}}$	1.979
	$[\phi_{E} \times \chi_{F_2}]_{F_2}$	$\frac{1}{2}(2\mu_u + \mu_d + \mu_c) + \mu_{\bar{c}}$	1.164
	$[\phi_{F_1} \times \chi_{F_2}]_{F_2}$	$\frac{1}{4}(6\mu_u + \mu_d + \mu_c) + \mu_{\bar{c}}$	2.232

diquark-diquark-antiquark model of pentaquark configurations. To the best of our knowledge, the present calculation is the first for a constituent quark model of pentaquarks.

The magnetic moment of a multiquark system is given by the sum of the magnetic moments of its constituent parts [34]

$$\vec{\mu} = \vec{\mu}_{\rm spin} + \vec{\mu}_{\rm orb} = \sum_i \mu_i (2\vec{s}_i + \vec{l}_i) \ , \tag{12}$$

where μ_i denotes the magnetic moment of the i−th quark or antiquark.

Since at present, we do not consider radial or orbital excitations, the magnetic moment only depends on the spin part. The results for the different $uudc\bar{c}$ configurations are given in Table 3. The numerical values are obtained by using

$$\begin{aligned} \mu_u &= +1.852\ \mu_N \ , \\ \mu_d &= -0.972\ \mu_N \ , \\ \mu_s &= -0.613\ \mu_N \ , \\ \mu_c &= +0.404\ \mu_N \ . \end{aligned} \tag{13}$$

The magnetic moments of the up, down and strange quarks were fitted to the magnetic moments of the proton, neutron and Λ hyperon. The value of the magnetic moment of the charm quark is consistent with a constituent mass of 1.550 GeV.

5. Summary and conclusions

In conclusion, in this contribution we discussed the classification of $uudc\bar{c}$ pentaquark states with angular momentum and parity $J^P = 3/2^-$. At present we did not include orbital excitations. The coupling between the four-quark states $uudc$ and the antiquark $\bar{c}$ was carried out at the level of the $SU(3)$ flavor group which results in a pure $uudc\bar{c}$ state. Couplings at the level of the $SU(8)$ spin-flavor group or the $SU(4)$ flavor group will lead to admixtures with $uudu\bar{u}$, $uudd\bar{d}$ and $uuds\bar{s}$ configurations [39].

Since for future studies of pentaquark states in photoproduction reactions it is important to have information on the electromagnetic couplings, we presented a first step by calculating the magnetic moments of all allowed $uudc\bar{c}$ states. The numerical values are positive and in the range from $\sim 1 - 3\ \mu_N$.

Finally, if indeed the pentaquark exists, there should be an entire multiplet of pentaquark states. The study of the structure of these multiplets and the mass spectrum is the subject of work in progress [39].

Acknowledgments
This work was supported in part by grant IN109017 from PAPIIT-DGAPA, UNAM and grant 251817 from CONACyT, Mexico.

References

[1] Aaij R *et al.* (LHCb Collaboration) 2015 *Phys. Rev. Lett.* **115** 072001
[2] Aaij R *et al.* (LHCb Collaboration) 2016 *Phys. Rev. Lett.* **117** 082002
[3] Aaij R *et al.* (LHCb Collaboration) 2016 *Phys. Rev. Lett.* **117** 082003
[4] Kubarovsky V and Voloshin M B 2015 *Phys. Rev. D* **92** 031502(R)
[5] Wang Q, Liu X H and Zhao Q 2015 *Phys. Rev. D* **92** 034022
[6] Karliner M and Rosner J L 2016 *Phys. Lett. B* **752** 329
[7] Hiller Blin A N, Fernández-Ramírez C, Jackura A, Mathieu V, Mokeev V I, Pilloni A and Szczepaniak A P 2016 *Phys. Rev. D* **94** 034002
[8] Fernández-Ramírez C, Hiller Blin A N and Pilloni A 2017 *arXiv:1703.06928*
[9] Meziani Z E *et al.* 2016 *arXiv:1609.00676*
[10] Wu J J, Molina R, Oset E and Zou B S 2010 *Phys. Rev. Lett.* **105** 232001
[11] Wu J J, Molina R, Oset E and Zou B S 2011 *Phys. Rev. C* **84** 015202
[12] Yang Z C, Sun Z F, He J, Liu X and Zhu S L 2012 *Chin. Phys. C* **36** 6
[13] Karliner M and Rosner J L 2015 *Phys. Rev. Lett.* **115** 122001
[14] Chen R, Liu X, Li X Q and Zhu S L 2015 *Phys. Rev. Lett.* **115** 132002
[15] Chen H X, Chen W, Liu X, Steele T G and Zhu S L 2015 *Phys. Rev. Lett.* **115** 172001
[16] Roca L, Nieves J and Oset E 2015 *Phys. Rev. D* **92** 094003
[17] He J 2016 *Phys. Lett. B* **753** 547
[18] Eides M I, Petrov V Yu, Polyakov M V 2016 *Phys. Rev. D* **93** 054039
[19] Yamaguchi Y and Santopinto E 2016 *arXiv:1606.08330*
[20] Maiani L, Polosa A D and Riquer V 2015 *Phys. Lett. B* **749** 289
[21] Lebed R F 2015 *Phys. Lett. B* **749** 454
[22] Wang G J, Chen R, Ma L, Liu X and Zhu S L 2016 *Phys. Rev. D* **94** 094018
[23] Yang G, Ping J and Wang F 2017 **Phys. Rev. D 95** 014010
[24] Deng C, Ping J, Huang H and Wang F 2017 *Phys. Rev. D* **95** 014031
[25] Takeuchi S and Takizawa M 2017 *Phys. Lett. B* **764** 254
[26] Santopinto E and Giachino A 2017 *arXiv:1604.03769v2*
[27] Guo F K, Meissner U G, Wang W and Yang Z 2015 *Phys. Rev. D* **92** 071502
[28] Liu X H, Wang Q and Zhao Q 2016 *Phys. Lett. B* **757** 231
[29] Mikhasenko M 2015 arXiv:1507.06552
[30] Burns T J 2015 *Eur. Phys. J. A* **51** 152
[31] Chen H X, Chen W, Liu X and Zhu S L 2016 *Phys. Rep.* **639** 1
[32] Patrignani C *et al.* (Particle Data Group) 2016 *Chin. Phys. C* **40** 100001
[33] Bijker R, Giannini M M and Santopinto E 2004 *Eur. Phys. J. A* **22** 319
[34] Bijker R, Giannini M M and Santopinto E 2004 *Phys. Lett. B* **595** 260
[35] Bjørken B J and Glashow S L 1964 *Phys. Lett.* **11** 255
[36] Glashow S L, Iliopoulos J and Maiani L 1970 *Phys. Rev. D* **2** 1285
[37] Gaillard M K, Lee B W and Rosner J L 1975 *Rev. Mod. Phys.* **47** 277
[38] De Rújula A, Georgi H and Glashow S L 1975 *Phys. Rev. D* **12** 147
[39] Bijker R and Fernández-Ramírez C 2017, work in progress

IOP Conf. Series: Journal of Physics: Conf. Series **876** (2017) 012005 doi:10.1088/1742-6596/876/1/012005

Neutron decay correlations in the Nab experiment

L. J. Broussard[1], R. Alarcon[2], S Baeßler[1,3], L. Barrón Palos[4], N. Birge[5], T. Bode[6], J. D. Bowman[1], T. Brunst[6], J. R. Calarco[7], J. Caylor[5], T. Chupp[8], V. Cianciolo[1], C. Crawford[9], G. W. Dodson[1], J. DuBois[3], W. Fan[3], W. Farrar[3], N. Fomin[5], E. Frlež[3], J. Fry[3], M. T. Gericke[10], F. Glück[11], G. L. Greene[1,5], R. K. Grzywacz[5], V. Gudkov[12], C. Hendrus[8], F. W. Hersman[7], T. Ito[13], H. Li[3], N. Macsai[10], M. F. Makela[13], J. Mammei[10], R. Mammei[14], J. Martin[14], M. Martinez[2], P. L. McGaughey[13], S. Mertens[6], J. Mirabal-Martinez[13], P. Mueller[1], S. A. Page[10], S. I. Penttilä[1], R. Picker[15], B. Plaster[9], D. Počanić[3], D. C. Radford[1], J. Ramsey[13], K. P. Rykaczewski[1], A. Salas-Bacci[3], E. M. Scott[5], S. K. L. Sjue[13], A. Smith[3], E. Smith[13], A. Sprow[9], E. Stevens[3], J. Wexler[16], R. Whitehead[5], W. S. Wilburn[13], A. R. Young[16], B. A. Zeck[13,16]

[1]Physics Division, Oak Ridge National Laboratory, Oak Ridge, TN 37831, USA
[2] Department of Physics, Arizona State University, Tempe, AZ 85287, USA
[3] Physics Department, University of Virginia, Charlottesville, VA 22904, USA
[4] Universidad Nacional Autónoma de México, México, D.F. 04510, MX
[5] Department of Physics and Astronomy, University of Tennessee, Knoxville, TN 37996, USA
[6] Max Planck Institute for Physics, Föhringer Ring 6, 80805 Munich, DE
[7] Department of Physics, University of New Hampshire, Durham, NH 03824, USA
[8] Department of Physics, University of Michigan, Ann Arbor, MI 48109, USA
[9] Department of Physics and Astronomy, University of Kentucky, Lexington, KY 40506, USA
[10] Department of Physics, University of Manitoba, Winnipeg, MB R3T 2N2, CA
[11] IEKP, Universität Karlsruhe (TH), Kaiserstraße 12, 76131 Karlsruhe, DE
[12] Department of Physics and Astronomy, University of South Carolina, Columbia, SC 29208, USA
[13] Physics Division, Los Alamos National Laboratory, Los Alamos, NM 87545, USA
[14] Department of Physics, University of Winnipeg, Winnipeg, MB R3B 2E9, CA
[15] TRIUMF, Vancouver, BC V6T 2A3, CA
[16] Physics Department, North Carolina State University, Raleigh, NC 27695, USA

E-mail: broussardlj@ornl.gov

Abstract.
The Nab experiment will measure the correlation a between the momenta of the beta particle and antineutrino in neutron decay as well as the Fierz term b which distorts the beta spectrum.

The a correlation can be used to extract the quark mixing matrix element V_{ud}, which is used to precisely test our understanding of the electroweak interaction. A non-zero Fierz term would signal the presence of scalar and tensor currents which must come from new physics beyond the Standard Model of Particle Physics. Nab uses an asymmetric spectrometer to determine the proton time-of-flight and the beta energy in order to reconstruct the angular correlation between the electron and antineutrino. The charged particles are detected using thick, large-area, segmented silicon detectors, which must be held at a potential of -30 kV to detect the very low energy protons. A 24 channel prototype system has been tested and found to meet specifications of 3 keV full-width half maximum (FWHM) energy resolution at 30 keV, 50 ns signal rise times, and thresholds < 10 keV. The 128 ch fully instrumented detection system is currently in development. Silicon drift detectors in development by the TRISTAN collaboration can reach an energy resolution as low as 135 eV FWHM at 6 keV, and could enable the next generation of neutron beta decay experiments.

1. Introduction

Precision measurements in neutron decay are used to test the weak interaction in the Standard Model of Particle Physics, and provide important inputs for astrophysical processes such as solar fusion cross sections [1] and big-bang nucleosynthesis calculations [2]. One of the most precise tests of the weak interaction is performed by testing the requirement that the Cabibbo Kobayashi Maskawa (CKM) matrix that describes the mixing of quarks in the weak interaction must be unitary. The uncertainty of the most demanding check of unitarity, $V_{ud}^2 + V_{us}^2 + V_{ub}^2 = 1$, is dominated by the matrix element $V_{ud} = 0.97417(21)$, currently determined most precisely from the lifetimes of the set of superallowed Fermi $0^+ \rightarrow 0^+$ decays [3]. Recently, updated lattice QCD simulations have introduced some tension into the values of V_{us} determined from kaon decays and therefore CKM unitarity [4]. This situation highlights the importance of having multiple independent systems as a check against unidentified experimental errors, or for possible new physics.

V_{ud} is also determined from several other systems but with much larger uncertainty: the $T = \frac{1}{2}$ mirror nuclear decays ($V_{ud} = 0.9719(17)$) [5], pion decay ($V_{ud} = 0.9728(30)$ [6], and neutron decay ($V_{ud} = 0.9758(16)$) [7]. Neutron decay is not sensitive to some of the nucleus-dependent theoretical corrections that are included in the superallowed Fermi and mirror nuclear decays, and with improved experimental uncertainties should allow for a more precise determination of V_{ud}. Because the neutron is a mixed Fermi and Gamow-Teller decay, two observables are required to fix the ratio of axial-vector to vector form factors λ, which must be experimentally determined, and extract V_{ud}.

To be competitive with superallowed $0^+ \rightarrow 0^+$ Fermi decays, the uncertainty in the neutron lifetime must be improved to ~0.3 s and the uncertainty in λ to roughly the 10^{-4} level. The Particle Data Group (PDG) recommendation for the neutron lifetime is 880.3(1.1) [7], however there is a ~3 σ discrepancy between experiments performed with cold neutrons [8, 9] and ultracold neutrons [10–14]. The parameter λ is known primarily from the angular correlation between the neutron spin and the emitted beta particle (the β-asymmetry A) from $A = \frac{-2|\lambda|(|\lambda|+1)}{1+3\lambda^2}$ in the Standard Model prediction. The most recent and precise measurements of the β-asymmetry are from the UCNA and Perkeo II experiments, which found $\lambda = -1.2755(30)$ [15] and $-1.2748(14)$ [16], respectively, and are somewhat in tension with previous measurements.

The correlation between the electron and antineutrino momenta, a, can be used to extract λ with nearly the same precision as A, from $a = \frac{1-\lambda^2}{1+3\lambda^2}$. The value of this correlation recommended by the PDG is a = -0.103(4). The most precise determinations of a to date have used the technique of measuring the proton energy spectrum distortion [17,18]. The slight negative value of a results in a slight preference for electrons and antineutrinos to be anti-aligned, and therefore an excess of protons with lower energies compared to higher energies, which is apparent in a measurement of the shape of the proton energy spectrum.

Two ongoing experimental efforts, aSPECT [19] and aCORN [20], are attempting to improve the precision in a by taking the approach of measuring the proton energy spectrum and determining the momenta of the proton and electron, respectively. The aSPECT experiment uses a retardation spectrometer to determine the integral of the proton energy spectrum, with a variable retardation voltage between the cold neutron beam and a silicon drift detector so that only protons with sufficient kinetic energy can pass to be detected. aSPECT is currently analyzing data from its 2013 run with an expected precision of $\delta a/a \sim 1\%$ [21], and an ultimate reach of $\delta a/a \sim 0.3\%$ is achievable.

The aCORN experiment instead determines a by comparing the ratio of counts with proton times-of-flight corresponding to aligned and anti-aligned electrons and antineutrinos. The aCORN spectrometer only accepts electrons and protons with momenta nearly parallel to the spectrometer axis, and electrons emitted in one direction. Therefore, the antineutrino momentum is nearly parallel with the electron momentum and has equal probability to be aligned or anti-aligned for accepted events, except for the influence of the parameter a. aCORN performed a data-taking run at the National Institutes of Standards and Technology NG-6 beamline from 2013-2014 and is currently analyzing the data. They expect to reach an uncertainty of a few percent. aCORN recently completed a second run with an improved apparatus on the high flux NG-C beamline in 2016 with a goal uncertainty of $\delta a/a \sim 1\%$.

2. Nab Experiment Overview

The aSPECT approach of measuring an asymmetry in proton time-of-flight allows for a more sensitive determination of a than by performing proton spectroscopy, but suffers from reduced statistics available due to the limited acceptance of the spectrometer. The Nab collaboration has proposed a different approach [22, 23] that also uses the proton time-of-flight to determine a, but with full kinematic acceptance of the electrons. The Nab experiment is the next major experiment to use the Fundamental Neutron Physics Beamline at the Spallation Neutron Source at Oak Ridge National Laboratory [24]. It is currently being installed and is expected to begin commissioning in early 2018. Nab is designed to achieve a precision in a of $\delta a/a \sim 10^{-3}$ as well as place a limit on a possible Fierz interference term which distorts the beta energy spectrum of $\delta b \sim 10^{-3}$. The Fierz interference term has never been measured in neutron decay and is predicted to be zero in the Standard Model. A non-zero Fierz term would indicate the presence of Scalar or Tensor currents which must originate from Beyond Standard Model physics [25,26]. Currently, very tight constraints on Scalar currents comes from the set of superallowed $0^+ \rightarrow 0^+$ decays [3], and Tensor currents are best constrained from pion decay [27, 28], or neutron and nuclear decay [29, 30].

We can understand how to extract the parameter a from the measurable quantities in Nab by considering conservation of momentum,

$$p_p^2 = p_e^2 + 2p_e p_\nu cos\theta_{e\nu} + p_\nu^2, \tag{1}$$

where the valid values of proton momentum p_p are given by the requirement that the cosine of the angle $\theta_{e\nu}$ between the electron and antineutrino momenta p_e and p_ν, respectively, be between ± 1. In the infinite nuclear mass approximation, both p_e and p_ν are a function of only E_e. The kinematically allowed values for the squared proton momentum as a function of E_e and cos $\theta_{e\nu}$ form a tear-drop shape (Fig. 1). At fixed E_e (assuming zero neutrino mass) the neutron decay rate W is proportional to

$$W \propto 1 + a\frac{p_e cos\theta_{e\nu}}{E_e} \tag{2}$$

IOP Conf. Series: Journal of Physics: Conf. Series **876** (2017) 012005 doi:10.1088/1742-6596/876/1/012005

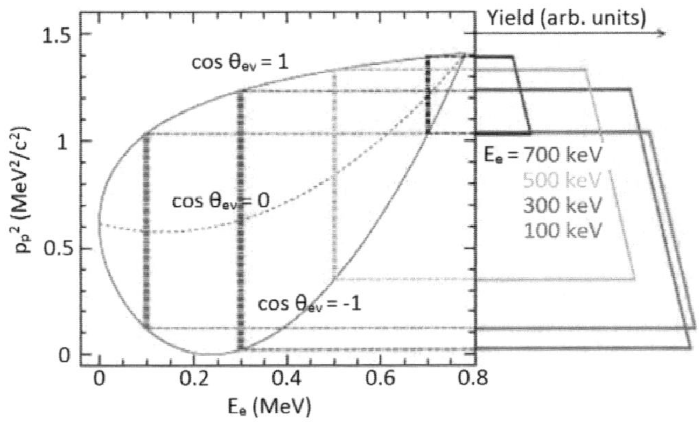

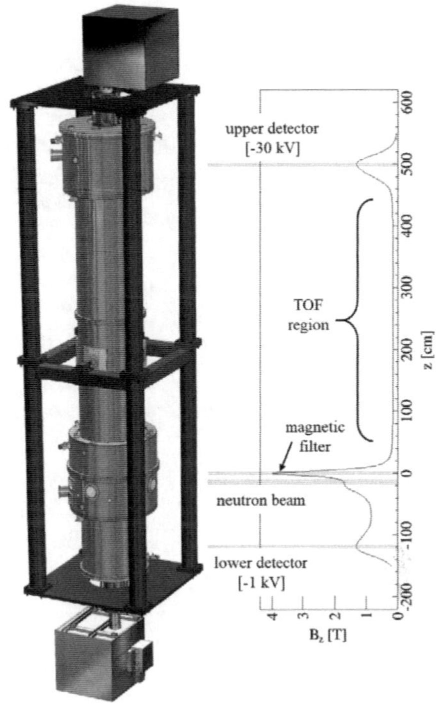

Figure 1. (Top) Allowed values of squared proton momentum as a function of electron energy. The slope of the yield vs. squared proton momentum at fixed electron energy is proportional to the electron-neutrino correlation, a. (Right) The Nab spectrometer and its on-axis magnetic field profile. The spectrometer is shown in green, the passive shield/mounting structure is purple, and the high voltage enclosure for the data acquisition system is gold (color online).

where $\cos\theta_{e\nu}$ depends linearly on p_p^2. Therefore, the slope of W vs. p_p^2 at fixed E_e is proportional to the parameter of interest, a.

The energy of the proton is quite small, with an endpoint of 751 eV, making the direct determination of the proton momentum quite challenging. In the magnetic spectrometer, the charged particles that result from the decay spiral along magnetic field lines and are guided to a detector, so that the time-of-flight depends more strongly on emission angle with respect to the spectrometer axis than on momentum. The strategy in Nab is to select only protons emitted in a narrow cone along the spectrometer axis, and then use a large field expansion to further parallelize the momentum. The Nab spectrometer and magnetic field profile are depicted in Fig. 1. The spectrometer is asymmetric, with a longer arm used for the proton time-of-flight measurement. Just above the neutron beam guide, which resides in a field of about 1.7 T, the magnetic field increases sharply to 4 T. This magnetic field "pinch" reflects protons that are not nearly parallel with the field lines. To longitudinalize the particles' momenta, the field then drops to 0.2 T for the length of the time-of-flight region. The detectors are positioned at about 1.3 T. Construction of the magnet is nearly complete, with an expected delivery during the summer of 2017.

Because of their low energy, protons require an accelerating potential to be detected. To measure the a correlation, only the upper detector has the accelerating potential applied to detect protons, whereas electrons are measured in either detector. A small accelerating potential is applied to the lower detector which is insufficient to allow proton detection but prevents the protons from being reflected back to the upper detector and skewing the time-of-flight measurement. To measure the b correlation, the -30 kV accelerating potential is applied to the lower detector where there is no field pinch, to increase the counting rate.

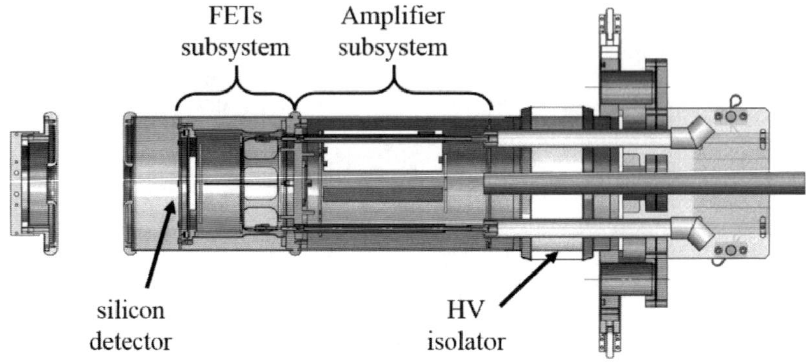

Figure 2. Prototype detector mounting assembly (color online).

3. Prototype Detection System

The detectors developed for Nab are unique in that they were designed to enable direct detection of both the recoil proton and electron with excellent energy and timing resolution and good position sensitivity. They are 1.5–2 mm thick, with ~100 cm^2 active area silicon detectors, segmented into 127 hexagonal pixels. The detectors were developed in collaboration with Micron Semiconductor, Ltd. [31] and the UCNB collaboration [32, 33]. The detectors exceed the requirements of the Nab experiment, with a 100 nm entrance window (dead layer) and 3 keV energy resolution measured using a proton beam at the Triangle Universities Nuclear Laboratory [34]. Protons with energies as low as 15 keV were detected. In Nab, a -30 kV accelerating potential will be used for proton detection.

A prototype detection system has been developed to characterize the expected performance (Fig. 2). The prototype includes 19 instrumented single pixels plus 18 pixels ganged into 4 channels of 4 or 5 pixels. The preamplifier consists of two subsystems, with the first containing 3 boards of 8 channels with the FETs and feedback loop, mounted at the detector in vacuum and cooled by connection to liquid nitrogen lines, and the second containing 4 boards of 6 channels with later gain stages, mounted in air at room temperature. The data acquisition system is based on the National Instruments PXIe-5171R digitizers [35]. The mounting structure was designed to mount into the UCNA spectrometer [36] and position the detectors in the field expansion region, at 0.6 T. The design requirements for the UCNA spectrometer are similar to that of Nab, in particular, the diameter is the same. The mount includes a 30 kV ceramic high voltage isolator, which allows a -30 kV accelerating potential to be applied to the detector, electronics, and data acquisition system.

The performance of the prototype detection system has been demonstrated to meet specifications for Nab [37]. The preamplifier signals have rise times of ~50 ns, which translates to a time resolution of several ns, depending on signal amplitude. The energy resolution was determined to be ~ 3 keV at 30 keV using X-ray sources, with an energy threshold < 10 keV. The leakage current per pixel was ~ 1 pA at the operating detector temperature of 175 K. The preamplifier electronics were found to exhibit no nonlinearities above 0.3%, the limit of the input pulser used for characterization.

Electron-proton coincidences from neutron beta decay were measured using this system. The measurement was performed using ultracold neutrons from the LANSCE source [38]. To increase the same-side coincidence rate the neutrons were polarized by a 6 T magnet followed by an adiabatic fast passage spin flipper [39]. Coincident protons are identified by their energy deposition in the active part of the Si detector of slightly less than 20 keV, and by their delayed

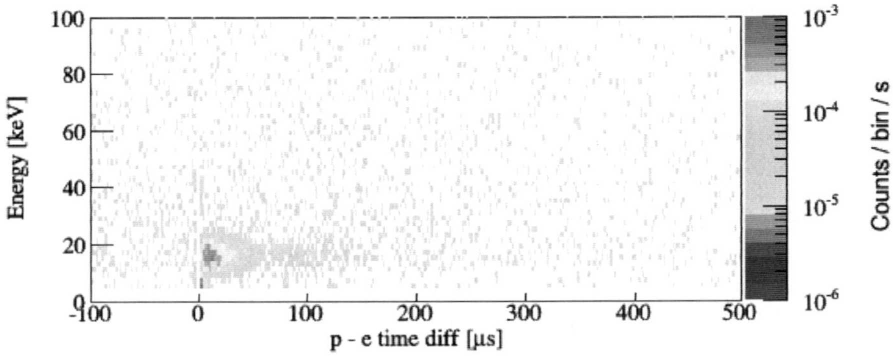

Figure 3. Protons deposit < 20 keV in the detector after the dead layer and arrive tens of μs after an initial electron event (color online).

detection time relative to the electron (Fig. 3). This system has also been used to demonstrate an important systematic consideration for experiments which use finite detectors in magnetic spectrometers, in a recent study of the non-monotonic point-spread function of monoenergetic particles in a homogeneous magnetic field [40].

4. Fully Instrumented Detection System

The design of the fully instrumented detection system includes several considerations to improve reliability and performance over the prototype. One major change is the interface to the detector. The prototype detectors included surface mounted pin header sockets for connection to the preamplifier assembly and bias input, which were easily damaged after many reconnections. The strategy chosen for the fully instrumented system was inspired by the KATRIN detection system [41], and instead uses spring-loaded pogo pins to make electrical contact. The concept was tested using a 4 channel prototype in which the preamplifier assembly directly contacted the gold contact of a 1 in. quadrant detector developed by Micron Semiconductor Ltd. The prototype was physically robust to dozens of reconnection and cryogenic cycles.

An alternate contact configuration was chosen for the detectors to further improve confidence that the detectors could not be damaged, and to meet other requirements of the detection system. In the chosen design, the pogo-pins contact pads on the ceramic mount of the detector, which are wire-bonded to the pixels (Fig. 4). All silicon pixels and partial pixels are accessible through the array of holes in the ceramic. The interface board is a Rogers 4003C ceramic composite [42], which includes the pogo-pins to contact the detector as well as contact pads on the back-side to interface to the preamplifier electronics, also through pogo-pins. This interface will serve as a vacuum feedthrough, such that the electronics can be disconnected without breaking vacuum or warming up the cold-bore spectrometer, avoiding several weeks of delay if an electronics channel should fail. A prototype interface board with through-hole pogo-pins and an indium vacuum seal was found to have a helium leak rate below 10^{-9} mbar-l/s at cryogenic temperature. The detectors are currently being assembled by Micron Semiconductor Ltd.

The fully instrumented system multiplies the number of channels by more than five, but the channel density of the boards is the same as in the prototype. This design implements the 16 FET boards and 22 amplifier boards radially about the spectrometer axis (Fig. 4). The FET assembly is mounted directly behind the detector and contacts the detector interface by pogo-pin contacts. It is mounted inside a copper cold can and resides in a separate vacuum from the detector and spectrometer. The FET assembly and amplifier assembly is separated by another

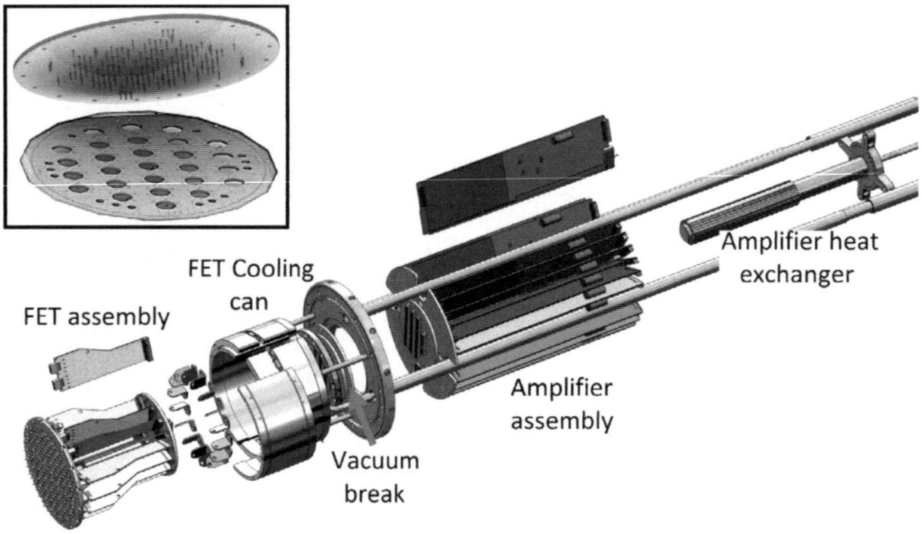

Figure 4. Exploded diagram of the 128 channel, fully instrumented preamplifier electronics. (Inset) Redesigned detector interface, including pogo-pin contacts (color online).

Rogers 4003C interface board which serves as a high channel density vacuum feedthrough. The amplifier assembly is mounted around a central axis which provides water cooling for the boards to maintain this section at room temperature.

The redesigned mounting structure includes an improved electrode design for better electric field uniformity and reduced Penning traps, especially near the ceramic high voltage break, which could lead to high voltage breakdown (Fig. 5). It is re-entrant into the Nab spectrometer to position the detectors at the 1.3 T region. The design allows for the preamplifier assembly to be extracted without venting the spectrometer, and for the amplifier assembly to be extracted separately. Provision for a load lock assembly is included, if a need to change detectors frequently is identified. In the prototype system, liquid nitrogen was used to cool the detector and FET assembly; however this was associated with added microphonics noise in the detector signals.

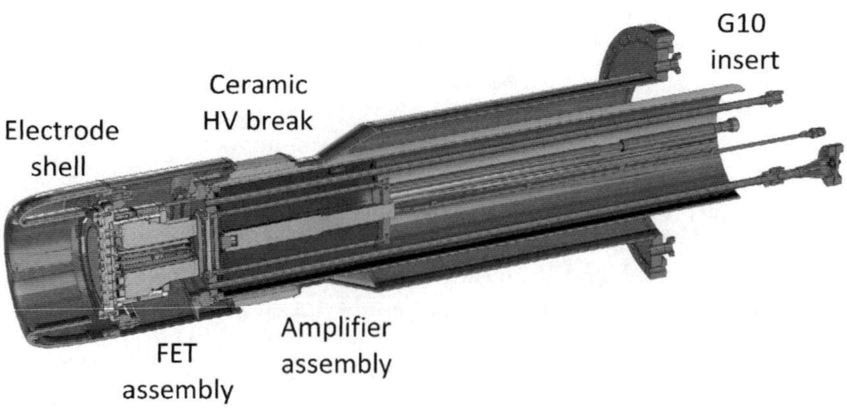

Figure 5. Diagram of mount with detector and electronics installed (color online).

The detector and FET assembly will instead be cooled by circulating cold helium gas in the cooling lines. As the FET assembly is in its own vacuum system, an exchange gas can be used to more rapidly cool the assembly. A copper-coated G10 insert isolates the amplifier assembly and cooling/vacuum lines from the mounting structure held at ground potential. The mounting assembly is currently in fabrication.

5. Detectors for Future Neutron Decay Experiments

The detectors developed for Nab represent a significant advance for measurements of neutron beta decay observables. To achieve even higher sensitivities in next generation experiments, for example ABba [43], further improvements in detector technology may be required. The TRISTAN project is developing detectors which may be of interest for future neutron beta decay experiments. TRISTAN will use the KATRIN spectrometer [44] to search for a signature of a keV-scale sterile neutrino in the beta spectrum of tritium decay, which appears as a "kink" in the shape of the spectrum [45]. The TRISTAN experiment will improve upon the silicon detectors currently used by KATRIN [41] by implementing silicon drift detectors in a hexagonal pixel array. The "point-contact" anodes reduce the capacitance and therefore noise of the pixel, even for large pixel area. An energy resolution as low as 135 eV (150 eV) FWHM at 6 keV has been achieved with prototype detectors with (and without) detector cooling and using a Cube CMOS preamplifier developed by XGLab [46] (Fig. 6).

A drawback of the silicon drift detector is the increased time for the electron-hole pairs to be collected. The drift paths of charges in various detector geometries were simulated using a modified version of the SIGGEN code [47] (Fig. 6). To achieve similar rise times as the Nab experiment (< 50 ns), pixel diameters could be reduced to $\sim$1 mm. To avoid substantially increasing the number of channels that must be instrumented, several pixel outputs could potentially be combined into a single electronics channel. The point-contact anodes are most relevant for applications that do not require fast timing, such as for detection of only protons. The silicon drift detector used by aSPECT has 1 cm^2 pads and has drift times of order μs. It has a 30 nm aluminum layer covering the silicon entrance window, which has a charge collection efficiency that slowly varies from 0% to 100%.

A detector with a very thin dead layer in a form factor similar to the Nab detectors would significantly reduce systematic uncertainties related to detection of the proton and electron. The dead layer introduces an uncertainty in the energy reconstruction of the electron energy and in identifying backscattered electrons which deposit only a portion of their energy in the detector and traverse the dead layer twice. The dead layer is also an important limitation in the detector's ability to detect protons. Fig. 6 depicts the required accelerating voltage to produce a detectable proton for several values of dead layer thickness and deposited energy (above the noise), as predicted by the SRIM code [48]. A dead layer as thin as 30 nm or even 10 nm of silicon may be achievable for a thick and large area similar to the dimensions required for Nab. Reducing the required accelerating potential would dramatically simplify the design constraints of the detection system mount in future experiments.

6. Summary

The novel approach used by the upcoming Nab experiment will enable a precision measurement of the electron neutrino correlation at the level of $\delta a/a \sim 10^{-3}$ and the first direct measurement of the Fierz interference term at the level of $\delta b \sim$ few 10^{-3} in neutron decay. A prototype detection system has been used to detect the proton and electron in coincidence from neutron beta decay with rise times of 50 ns, energy resolution of 3 keV, and thresholds < 10 keV. The fully instrumented, 128 ch detection system is in fabrication. The Nab experiment is currently being installed at the Fundamental Neutron Physics Beamline at the Spallation Neutron Source

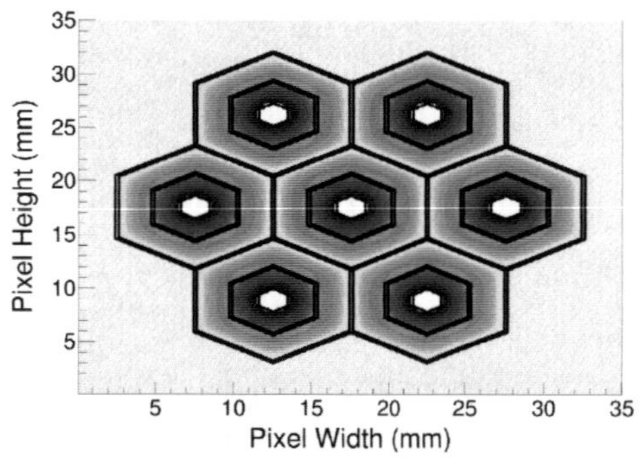

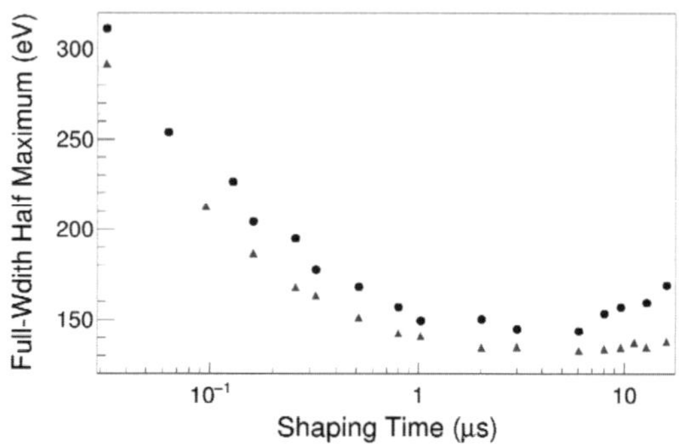

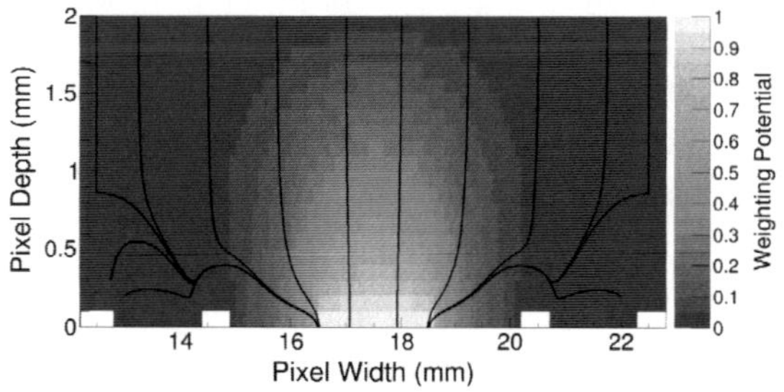

Silicon dead layer	Energy after dead layer (keV) 15 keV	10 keV	5 keV
100 nm	26	20	13
50 nm	20	14	8
30 nm	18	12	7
10 nm	16	11	6

Figure 6. (Top left) Schematic of hexagonal silicon drift detector with dimensions similar to the Nab detector. The steering rings (black) surrounding each anode (white) speed up charge collection time and prevent areas of incomplete charge collection. (Top right) Full-width half maximum of ~6 keV X-rays from ^{55}Fe vs. shaping time for detector uncooled (black circle) and cooled to -30° C (blue triangle). (Bottom left) Weighting potential of central pixel and simulated drift paths of charges in a detector with dimensions similar to the Nab detector. Steering electrodes are indicated by white and the anode by yellow. (Bottom right) Required accelerating potential for protons to deposit the indicated energy in the active area of the detector for various values of a silicon dead layer. (color online)

with expected commissioning in 2018. Future detector designs with thinner dead layer and point contact geometry appear promising and may be interesting for future neutron decay experiments.

Acknowledgments

Research was sponsored by the Laboratory Directed Research and Development Program of Oak Ridge National Laboratory, managed by UT-Battelle, LLC, for the U. S. Department of Energy [project number 8215], the LDRD program of Los Alamos National Laboratory [project number 20110043DR], the National Science Foundation [contract number NSF PHY-1205833, 1307426], and the U.S. Department of Energy, Office of Science, Office of Nuclear Physics [contract numbers DE-AC52-06NA25396 under proposal 2017LANLEEDM, DE-AC05-00OR22725, DE-FG02-03ER41258, DE-FG02-97ER41042, DE-SC0008107, DE-SC0014622].

References

[1] E. G. Adelberger *et al.* 2011 *Rev. Mod. Phys.* **83** 195
[2] R. H. Cyburt, B. D. Fields, K. A. Olive and T.-H. Yeh 2016 *Rev. Mod. Phys.* **88** 015004
[3] J. C. Hardy and I. S. Towner 2015 *Phys. Rev. C* **91** 025501
[4] S. Aoki *et al.* (Flavour Lattice Averaging Group) 2017 *Eur. Phys. J. C* **77** 112
[5] O. Naviliat-Cuncic and N. Severijns 2009 *Phys. Rev. Lett.* **102** 142302
[6] D. Počanić *et al.* 2004 *Phys. Rev. Lett.* **93** 181803

[7] C. Patrignani *et al.* (Particle Data Group) 2016 *Chin. Phys. C* **40** 100001
[8] J. Byrne, P. G. Dawber, C. G. Habeck, S. J. Smidt, J. A. Spain and A. P. Williams 1996 *Eur. Phys. Lett.* **33** 187
[9] A. T. Yue, M. S. Dewey, D. M. Gilliam, G. L. Greene, A. B. Laptev, J. S. Nico, W. M. Snow, and F. E. Wietfeldt 2013 *Phys. Rev. Lett.* **111** 222501
[10] W. Mampe, L. N. Bondarenko, V. I. Morozov, Yu. N. Panin. and A. I. Fomin 1993 *J. Exp. Theor. Phys.* **57** 82 (translated from *Pis'ma Zh. Eksp. Teor. Fiz.* **57** 77)
[11] A. Serebrov *et al.* 2005 *Phys. Lett. B* **605** 72
[12] A. Pichlmaier, V. Varlamov, K. Schreckenbach and P. Geltenbort 2010 *Phys. Lett. B* **693** 221
[13] A. Steyerl, J. M. Pendlebury, C. Kaufman, S. S. Malik and A. M. Desai 2012 *Phys. Rev. C* **85** 065503
[14] S. Arzumanov, L. Bondarenko, S. Chernyavsky, P. Geltenbort, V. Morozov, V. V. Nesvizhevsky, Yu. Panin and A. Strepetov 2015 *Phys. Lett. B* **745** 79
[15] M. P. Mendenhall *et al.* (UCNA Collaboration) 2013 *Phys. Rev. C* **87** 032501(R)
[16] D. Mund, B. Märkisch, M. Deissenroth, J. Krempel, M. Schumann, H. Abele, A. Petoukhov and T. Soldner 2013 *Phys. Rev. Lett.* **110** 172502
[17] C. Stratowa, R. Dobrozemsky and P. Weinzierl 1978 *Phys. Rev. D* **18** 3970
[18] J. Byrne, P. G. Dawber, M. G. D. van der Grinten, C. G. Habeck, F. Shaikh, J. A. Spain, R. D. Scott, C. A. Baker, K. Green and O. Zimmer 2002 *J. Phys. G* **28** 1325
[19] M Simson *et al.* 2009 *Nucl. Instr. Meth. Phys. Res. A* **611** 203
[20] F.E. Wietfeldt *et al.* 2009 *Nucl. Instr. Meth. Phys. Res. A* **611** 207
[21] R. Maisonobe *et al.* 2015 in "Measurement of the electron-antineutrino angular correlation coefficient a in neutron β-decay with the spectrometer aSPECT", in proceedings of "The European Physical Society Conference on High Energy Physics", PoS(EPS-HEP2015)595
[22] D. Počanić *et al.* 2009 *Nucl. Instr. Meth. Phys. Res. A* **611** 211
[23] S. Baeßler *et al.* 2013 *AIP Conf. Proc.* **1560** 114
[24] N. Fomin, G. L. Greene, R. R. Allen, V. Cianciolo, C. Crawford, T. M. Tito, P. R. Huffman, E. B. Iverson, R. Mahurin, W. M. Snow 2015 *Nucl. Instr. Meth. Phys. Res. A* **773** 45
[25] T. Bhattacharya, V. Cirigliano, S. D. Cohen, A. Filipuzzi, M. González-Alonso, M. L. Graesser, R. Gupta and H.-W. Lin 2012 *Phys. Rev. D* **85** 054512
[26] S. Gardner and B. Plaster 2013 *Phys. Rev. C* **87** 065504
[27] M. Bychkov *et al.* 2009 *Phys. Rev. Lett.* **103** 051802
[28] V. Cirigliano and S. Gardner and B. R. Holstein 2013 *Prog. Part. Nucl. Phys.* **71** 93
[29] F. Wauters, A. García and R. and Hong 2014 *Phys. Rev. C* **89** 025501, Erratum-ibid. 2015 **91** 049904
[30] R. W. Pattie, K. P. Hickerson and A. R. Young 2013 *Phys. Rev. C* **88** 048501, Erratum-ibid. 2015 **92** 069902
[31] Micron Semiconductor Ltd. `http://micronsemiconductor.co.uk` *(Accessed 3/1/2017)*
[32] W. S. Wilburn *et al.* 2009 *Rev. Mex. Fís. S* **55** 119
[33] L. J. Broussard 2012 *AIP Conf. Proc.* **1560** 149
[34] A. Salas-Bacci *et al.* 2014 *Nucl. Instr. Meth. Phys. Res. A* **735** 408
[35] National Instruments `http://www.ni.com/en-us/support/model.pxie-5171.html` *(Accessed 3/1/2017)*
[36] B. Plaster *et al.* 2012 *Phys. Rev. C* **86** 055501
[37] L. J. Broussard *et al.* 2017 *Nucl. Instr. Meth. Phys. Res. A* **849** 83
[38] A. Saunders *et al.* 2013 *Rev. Sci. Instrum.* **84** 013304
[39] A. T. Holley *et al.* 2012 *Rev. Sci. Instrum.* **83** 073505
[40] S. K. L. Sjue, L. J. Broussard, M. Makela, P. L. McGaughey, A. R. Young and B. A. Zeck 2015 *Rev. Sci. Instrum.* **86** 023102
[41] J. F. Amsbaugh *et al.* 2015 *Nucl. Instr. Meth. Phys. Res. A* **778** 40
[42] Rogers Corporation `http://www.rogerscorp.com/acs/products/54/ro4003c-laminates.aspx` *(Accessed 3/1/2017)*
[43] L. Barrón Palos *et al.* 2010 *J. Phys.: Conf. Series* **239** 012013
[44] J. Angrik *et al.* (KATRIN Collaboration) 2004 KATRIN design report, Wissenschaftliche Berichte FZKA 7090 `http://www.katrin.kit.edu/publikationen/DesignReport2004-12Jan2005.pdf`
[45] S. Mertens, T. Lasserre, S. Groh, G. Drexlin, F. Glück, A. Huber, A. W. P. Poon, M. Steidl, N. Steinbrink and C. Weinheimer 2015 *J. Cosmol. Astropart. Phys.* **2015** 20
[46] XGLab S. R. L. `www.xglab.it/low-noise-preamplifier.shtml` *(Accessed 3/1/2017)*
[47] D. Radford 2017 SIGGEN `http://radware.phy.ornl.gov/gretina/siggen`
[48] J. F. Ziegler and J. P. Biersack and M. D. Ziegler 2013 SRIM-2013 `www.srim.org`

IOP Conf. Series: Journal of Physics: Conf. Series **876** (2017) 012006 doi:10.1088/1742-6596/876/1/012006

Investigation of the Hoyle state in ^{12}C with a new hodoscope detector

D Dell'Aquila[1,2,3], I Lombardo[1,2], M Vigilante[1,2], M De Luca[1], L Acosta[4], C Agodi[5], F Cappuzzello[6,5], D Carbone[5], M Cavallaro[5], S Cherubini[5,6], A Cvetinovic[5], G D'Agata[6,5], L Francalanza[2], G L Guardo[5], M Gulino[7,5], I Indelicato[5], M La Cognata[5], L Lamia[6], A Ordine[2], R Pizzone[5], S Puglia[5], G Rapisarda[5], S Romano[5], G Santagati[5], R Spartà[5], C Spitaleri[6,5], A Tumino[7,5] and G Verde[3,8]

[1] Dip. di Fisica "E. Pancini", Università di Napoli Federico II, I-80126 Napoli, Italy
[2] INFN-Sezione di Napoli, I-80126 Napoli, Italy
[3] Institut de Physique Nuclaire, CNRS-IN2P3, Univ. Paris-Sud, Universit Paris-Saclay, 91406 Orsay Cedex, France
[4] National Autonomous University of Mexico, A.P. 20-364, México D.F. 01000 México
[5] INFN - Laboratori Nazionali del Sud, Via S. Sofia, I-95125 Catania, Italy
[6] Dip. di Fisica e Astronomia, Università di Catania, Via S. Sofia, I-95125 Catania, Italy
[7] Facoltà di Ingegneria ed Architettura, Università Kore, I-94100 Enna, Italy
[8] INFN - Sezione di Catania, Via S. Sofia, I-95125 Catania, Italy

E-mail: dellaquila@na.infn.it

Abstract. The 0_2^+ state in ^{12}C (7.654MeV, the Hoyle state) is important for the understanding of clustering phenomena in nuclei. The pronounced cluster nature of this state allows the triple-α process in stars with a reaction rate regulated by its structure properties. To precisely estimate the direct component in the 3α decay mechanism of the Hoyle state, we developed a new experiment using the ^{14}N(d,α)^{12}C reaction at 10.5MeV. An anti-coincidence telescope was used to identify the α ejectiles leading the residual ^{12}C in the Hoyle state, while its decays in 3α were studied by means of a new hodoscope of silicon detectors, superOSCAR, placed in kinematical coincidence to fully reconstruct the events. Details of the experiment and preliminary results are discussed in the text.

1. Introduction

Carbon isotopes represent a unique opportunity to characterize the properties of clustering phenomena as a function of their neutron-richness [1–9]. Indeed, several excited states based on molecular chains of α particles can be predicted in carbon neutron-rich isotopes starting from the assumption of a pronounced cluster nature for the Hoyle state in ^{12}C [10, 11].

The Hoyle state (7.654MeV, 0^+) is a well known cluster state of the self-conjugated ^{12}C isotope. Its pronounced cluster nature plays a fundamental role in Astrophysics, where it is involved in the so-called 3α process. This stellar process, which occurs in the helium-burning stage, regulates the nucleo-synthesis of carbon and heavier elements in the Universe. In 1952, it was indeed proposed by Salpeter [12] and Opik [13] that the synthesis of carbon in the red giant stars would proceed in a two-steps process, with the non-resonant radiative capture of

IOP Conf. Series: Journal of Physics: Conf. Series **876** (2017) 012006 doi:10.1088/1742-6596/876/1/012006

an α particle by the unbound ground state of ^{8}Be, produced in a previous α-α fusion reaction. Anyway, the short life time of ^{8}Be ($\approx 10^{-16}s$) limited the reaction rate of this process, making not possible to explain the amount of carbon and heavier elements present in the Universe. In 1953, Hoyle suggested that the presence of an s-wave resonance in the α+^{8}Be capture process, close to the corresponding separation energy, would increase the reaction rate of the production of ^{12}C [14], introducing what we now know as the *Hoyle state*. The structure of the Hoyle state is quite unusual and represents one of the most important unanswered questions in Nuclear Physics. Inelastic scattering experiments have pointed out a relatively large radius or this state, compared to the ground state of ^{12}C [15]. On a theoretical point of view, different hypothesis were formulated. As an example, while microscopic cluster model calculations describe the Hoyle state as a gas-like diluted structure [16], it appears as a bent-arm structure from *ab initio* calculations [17, 18] and a possible linear chain configuration was also predicted [15, 16]. All these properties are crucial in the determination of the 3α reaction rate, since it is strongly related to the corresponding decay partial widths.

The 3α reaction rate can be calculated by knowing the radiative and α decay widths of the Hoyle state. Radiative decay branching ratios are known to be a small amount of the total width ($\Gamma_{rad}/\Gamma = 4.12(11) \times 10^{-4}$ [19]), while, in good approximation, it is considered $\Gamma \approx \Gamma_{\alpha_0}$, where Γ_{α_0} represents the decay through α+^{8}Be (sequential decay), without contribution due to the corresponding direct decay in 3α ($\Gamma_{3\alpha}$). This assumption is well supported by [20], where an upper limit of $\Gamma_{3\alpha}/\Gamma < 4\%$ was given. Anyway, in recent times, after the results of Raduta et al. [21], that suggested a significantly larger value of this decay width ($\Gamma_{3\alpha}/\Gamma = 17.0\% \pm 5.0\%$), a number of high precision experiments were performed. All these experiments contradict the results of [21], but are not able to converge on a definitive conclusion. More in detail, while Kirsebom et al. [22] and Itoh et al. [23] do not evidence any direct component in the decay of Hoyle state, giving, respectively, the two upper limits $\Gamma_{3\alpha}/\Gamma < 0.5\%$ (C.L. 0.95) and $\Gamma_{3\alpha}/\Gamma < 0.2\%$ (C.L. 0.95), Rana et al. [24] and Morelli et al. [25] estimated a non-zero value of this branch ($\Gamma_{3\alpha}/\Gamma = 0.91\% \pm 0.14\%$ and $\Gamma_{3\alpha}/\Gamma = 1.1\% \pm 0.4\%$). The possible existence of a direct component in the 3α decay of Hoyle would result in a sizable modification of the triple-alpha reaction rate, especially in low temperature regions ($T_9 < 0.1$) of red giant stars, where the s-wave 2-step process does not dominate and possible direct processes could occur, increasing, as a consequence, the expected reaction rate [26, 27].

We report preliminary results on a new measurement of the 3α decay of the Hoyle state. We used the ^{14}N(d,α)^{12}C reaction at 10.5MeV with a backward anti-coincidence telescope, as discussed in [28], placed at $\theta_{lab} = 125°$, to identify α particles from the (d,α_2) reaction. The corresponding recoil is thus a ^{12}C excited in the Hoyle state, whose decay into 3α is studied with a new hodoscope of silicon detectors, superOSCAR. This paper is organized as follows: in the Section 2 details of the experiment are discussed, the Section 3 is dedicated to a preliminary description of the data analysis, while conclusions and future perspectives will be discussed in the Section 4.

2. Experimental details: the HOYLE experiment

The HOYLE (Hodoscope Oriented Yield Loader Experiment) experiment at INFN-Laboratori Nazionali del Sud (LNS, Catania, Italy) was developed with the aim of shedding light on the direct decay of the Hoyle state with large statistics and low background levels. The basic idea of this experiment is to use of the ^{14}N(d,α)^{12}C reaction to produce carbon recoils excited in the Hoyle state, and to fully reconstruct the 3α decays using an hodoscope of silicon detectors, superOSCAR. A 10.5MeV deuteron beam was provided by the 15MV tandem accelerator of the INFN-LNS with intensities not exceeding 4enA to avoid spurious coincidences. Deuterons induced reactions on a melamine target ($40\mu g/cm^2$, $C_3H_6N_6$) deposited on a carbon backing ($10\mu g/cm^2$). We used an anti-coincidence backward telescope to give the trigger for the

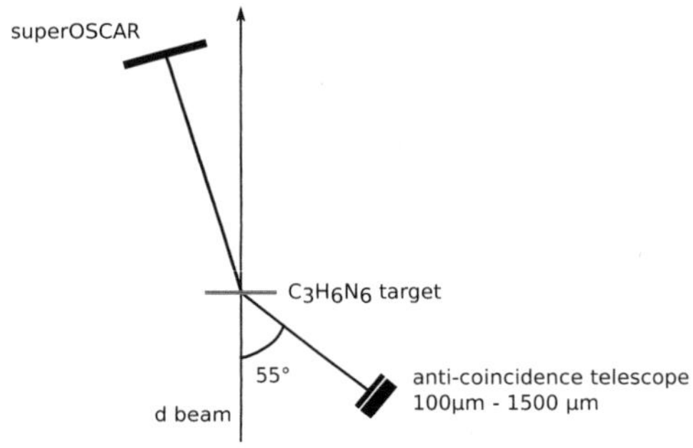

Figure 1. A scheme of the experimental device used in the HOYLE experiment. The anti-coincidence telescope is placed at 125° respect to the beam axis, while the superOSCAR hodoscope is at forward angles ($\approx 30°$).

experiment. The logic was the same as seen in [28]: a first stage is dimensioned to stop the α particles from the (d,α) groups, while the more energetic elastic deuterons and protons from (d,p) reactions, which appear as contaminants in the spectrum, punch through the first stage and give signal in the second stage. These particles can be removed from the trigger by using a fast anti-coincidence circuit between signals from the first and second stage. In other words, the anti-telescope gives a trigger to the experiment only if a particle is detected in the first stage without any signal in the second stage. With this technique it is possible to strongly reduce the contribution of contaminants to the experiment trigger. The used telescope detector is a two stages silicon 100μm-1500μm detector as indicated in Figure 1. The superOSCAR hodoscope is placed in kinematical coincidence with the center at 30° respect to the beam axis. It is constituted by 64 independent silicon pads (300μm thick), with energy resolutions of the order of 0.3% and active area of 1cm $\times$ 1cm, placed in 8×8 configuration. They are connected, with short SCI-SCI cables, to 4 pre-amplifier NPA-16FE, manufactured by the NeT Instruments and then to 4×16 channels NIM amplifiers. The advantage of using hodoscope detectors is that they provide better energy resolution than strip detectors and that they avoid the problem of tracks identification typical of the use of strips, as discussed in [23]. With this features we expect to strongly reduce the background level observed in previous works, where strips are used.

The energy released by the detected particles in the first stage of the backward telescope, which gives the trigger to the experiment, can be used in order to tag the excitation energy of the corresponding recoiling nucleus. The energy spectrum taken at 125° is reported with the blue line in the Figure 2, using a logarithmic scale. Peaks are well visible and they correspond to excited states populated in the ^{12}C residual nucleus or to deuteron-induced reactions on contaminants. The three high energy peaks correspond, respectively, to α particles leading the ^{12}C in the 9.63MeV, 7.654MeV and 4.44MeV excitation energy states. A continuous background is present under the identified peaks. Interestingly, by selecting only events for which we are able to reconstruct three particles inside superOSCAR, the blue spectrum becomes the green filled one (of Figure 2). This spectrum has a reduced statistics compared to the single particle spectrum, because of the geometrical efficiency of superOSCAR in the reconstruction of three particles, but the background results strongly reduced and almost vanishing in correspondence to the tails of the Hoyle peak. The green spectrum represents the backward spectrum obtained in case of fully reconstructed events, and, therefore, can be used to identify the class of events for which we will perform the analysis. In the following, only fully reconstructed events under the Hoyle peak of Figure 2 will be considered.

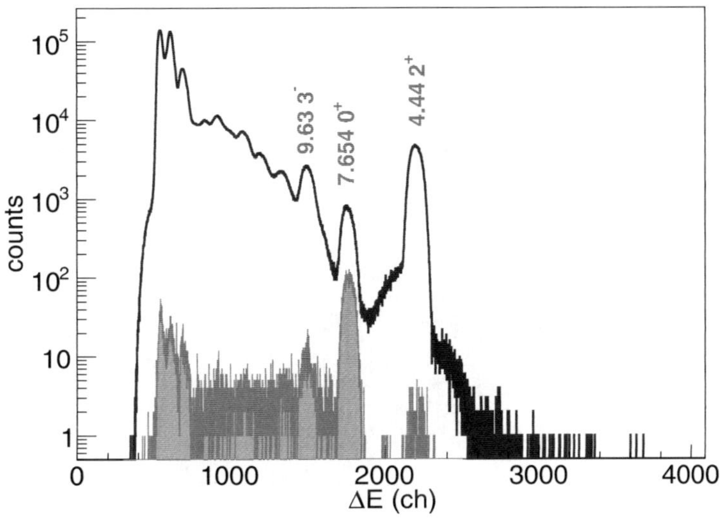

Figure 2. (Blue spectrum) Energy spectrum of particles detected with the anti-telescope technique at $\theta_{lab} = 125°$. States of the recoiling carbon close to the one of our interest are indicated by labels. (Green spectrum) energy spectrum obtained for fully reconstructed events. Only a part of the statistics was used to produce this Figure.

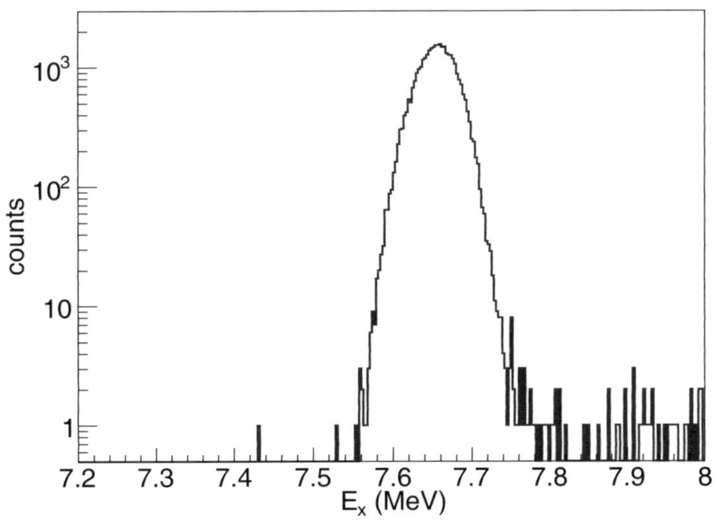

Figure 3. Invariant mass ^{12}C spectrum (E_x) obtained using fully reconstructed events, as shown in the Figure 2, and by selecting the Hoyle peak. A cut on the peak observed in the Q-value spectrum has been also done close to the Q_{gggg}.

3. Preliminary analysis

As a preliminary analysis, we will comment here on the possibility of correctly reconstruct the decay in 3α with our apparatus. After selecting the fully reconstructed events from the anti-

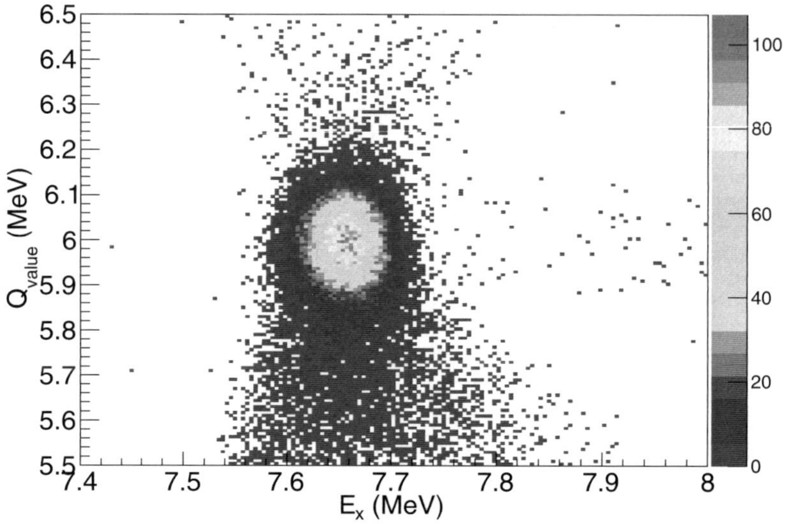

Figure 4. Bi-dimensional plot Q-value vs E_x. A peak is well visible indicating events in which both Q-value and invariant mass are well reconstructed.

coincidence telescope, we can proceed with the analysis of the 4 particles detected. A further improvement to the spectrum of Figure 2, and to the analysis itself, can be indeed obtained by analysing the detected particles in terms of Q-value and *invariant mass*. An inspection on the Q-value is the first step of the analysis, since it can be reconstructed without assumptions on the flight directions of the particles, using only kinetic energies. As expected, we find a prominent peak corresponding to the Q_{gggg} value, indicating the events correctly reconstructed. A simple selection on the Q-value spectrum allows to obtain, using the invariant mass technique on the 3α detected at forward angles, the excitation energy spectrum of the emitting ^{12}C, shown in Figure 3. The result is extremely satisfactory, the good energy resolution and angular segmentation of the superOSCAR device allow to reconstruct a narrow excitation energy peak ($FWHM \approx 50$keV) centered at 7.654MeV with a precision of 1keV. Furthermore, a very low background level is present under the peak, as testified by the right and left tails of the peak.

An improvement of the *fully reconstructed* class of events is finally possible by simultaneously selecting on Q-value and *invariant mass*. This is possible by using the bi-dimensional plot of Figure 4, where the reconstructed Q-value is plotted versus the excitation energy (Figure 3). The presence of a peak, with a low background level (see the color scale) allows to improve the selection by using circular cuts.

4. Conclusions and Perspectives

In conclusion, strong efforts were devoted to a precise measurement of the direct decay branching ratio of the Hoyle state in ^{12}C. We developed a new experiment at INFN-LNS by using the ^{14}N(d,α)^{12}C at 10.5MeV. A backward angle telescope detector was used to provide identification of the (d,α_2) contribution, strongly suppressing the background due to contaminant reactions by means of the anti-coincidence technique. We have shown that the corresponding decays in 3α of the Hoyle state can be precisely, and with a negligible background, reconstructed with the superOSCAR device. With the high resolution and unambiguous track identification of this device, we expect to give very soon further results in the resolution of such a complicated

IOP Conf. Series: Journal of Physics: Conf. Series **876** (2017) 012006 doi:10.1088/1742-6596/876/1/012006

problem.

References

[1] Spitaleri C *et al.* 2014 *Phys. Rev.* **C90** 035801 (*Preprint* 1407.4678)
[2] Lombardo I *et al.* 2016 *J. Phys. G.* **43** 45109
[3] Lombardo I *et al.* 2013 *Nucl. Instrum. Meth. Phys. Res.* **B 302** 19–23
[4] Lombardo I *et al.* 2014 *J. Phys. Conf. Ser.* **569** 012068
[5] Fritsch A *et al.* 2016 *Phys. Rev.* **C93** 014321
[6] Dell'Aquila D *et al.* 2016 *Phys. Rev. C* **93** 024611
[7] Cappuzzello F *et al.* 2012 *Phys. Lett.* **B711** 347–352
[8] Cappuzzello F *et al.* 2015 *Nat. Commun.* **6** 6743
[9] Cavallaro M *et al.* 2016 *Phys. Rev.* **C93** 064323
[10] von Oertzen W 1997 *Zeit. Phys.* **A 357** 355
[11] von Oertzen W, Freer M and Kanada-En'yo Y 2006 *Phys. Rep.* **432** 43–113
[12] Salpeter E 1952 *Phys. Rev.* **88** 547
[13] Opik E 1951 *Proc. R. Irish Acad. A* **54** 49
[14] Hoyle F 1954 *Astrophys. J. Suppl. Ser.* **1** 121
[15] Kamimura M 1981 *Nucl. Phys. A* **351** 456
[16] Uegaki E, Okabe S, Abe Y and Tanaka H 1977 *Prog. Theor. Phys.* **57** 1262
[17] Epelbaum E *et al.* 2011 *Phys. Rev. Lett.* **106** 192501
[18] Epelbaum E *et al.* 2012 *Phys. Rev. Lett.* **109** 252501
[19] Markham R, Austin S and Shahabuddin M 1976 *Nucl. Phys. A* **270** 489
[20] Freer M *et al.* 1994 *Phys. Rev. C* **49** R1751
[21] Raduta A *et al.* 2011 *Phys. Lett. B* **705** 65
[22] Kirsebom O *et al.* 2012 *Phys. Rev. Lett.* **108** 202501
[23] Itoh M *et al.* 2014 *Phys. Rev. Lett.* **113** 102501
[24] Rana T *et al.* 2013 *Phys. Rev. C* **88** 021601(R)
[25] Morelli L *et al.* 2016 *J. Phys.* **G43** 045110
[26] Ogata K, Kan M and Kamimura M 2009 *Prog. Theor. Phys.* **122** 1055
[27] de Diego R, Garrido E, Fedorov D and Jensen A 2011 *Phys. Lett. B* **695** 324
[28] Koenig W *et al.* 1977 *Il Nuov. Cim.* **39** 9

IOP Conf. Series: Journal of Physics: Conf. Series **876** (2017) 012007 doi:10.1088/1742-6596/876/1/012007

Pentaquark photoproduction

César Fernández-Ramírez

Instituto de Ciencias Nucleares, Universidad Nacional Autónoma de México, Ciudad de México 04510, Mexico

E-mail: cesar.fernandez@nucleares.unam.mx

Astrid N. Hiller Bin

Departamento de Física Teórica and IFIC, Centro Mixto Universidad de Valencia-CSIC, Institutos de Investigación de Paterna, E-46071 Valencia, Spain

Alessandro Pilloni

Theory Center, Thomas Jefferson National Accelerator Facility, Newport News, VA 23606, USA

Abstract. We present results and suggestions on how to confirm the existence and resonant nature of the $Pc(4450)$ detected at LHCb through photoproduction experiments. We find that this narrow structure might have escaped detection in past experiments and use those to give a constraint for the upper limit of the branching ratio/coupling to the $J/\psi\, p$ channel.

1. Introduction

One of the most remarkable properties of Quantum Chromodynamics (QCD) is the color confinement phenomenon that makes hadrons to be built out of color singlets of quarks. The simplest way to build a color neutral baryon (half-integer spin) state is out of three quarks (qqq). However, QCD tells us that there are other ways to build color neutral baryons, for example: hybrid baryons with gluonic components ($qqqg$, $qqqgg$, ...), pentaquarks ($qqqq\bar{q}$), heptaquarks ($qqqqq\bar{q}\bar{q}$) and so on. Finding complete multiplets of baryons beyond the qqq picture is one of the main goals of hadron spectroscopy and will provide insight on the dynamics of QCD in the hadronic energy regime. The recent discovery of two structures compatible with pentaquark states in the $\Lambda_b^0 \to J/\psi K^- p$ decay by LHCb [1] together with the observation of a compatible structure in $\Lambda_b^0 \to J/\psi \pi^- p$ [2], has triggered a lot of research activities aimed to explain them. Each signal has been associated to a pentaquark candidate, $P_c(4380)$ and $P_c(4450)$, with opposite parity assignments, and one state having spin 3/2 and the other 5/2 [1, 3]. The $P_c(4380)$ signal is less apparent and the properties of this state (mass and width) could depend on the amplitude analysis performed, especially for the K^-p background [3, 4]. However, the $P_c(4450)$ signal is very narrow and its mass and width are not expected to change by refinements in the amplitude analysis. Hence, its study has been the focus of many research groups. Several explanations for this signal have been proposed: a kinematical effect [5], a molecular state [6] and a compact QCD (pentaquark) state [7].

One of the best approaches to gain insight on the nature of $P_c(4450)$ consists on detecting the state in a different reaction. In [8] it was proposed to use $J/\psi\, p$ photoproduction. From the LHCb Λ_b^0 decay data we know that the $P_c(4450)$ state couples strongly to the $J/\psi\, p$, and the photon energy can be tuned to excite this particular state. This experiment has just been approved to run at Hall C in Jefferson Lab [9] and results are expected in the next few years. Hence, as a first step it is necessary to develop the amplitudes that will be needed to provide realistic predictions for the planning of the experiment and later to analyze the experimental data. In the next section we sketch the main features of the J/ψ photoproduction model that we developed with these purposes in mind. The detailed description of the model and the corresponding results can be found in [10].

2. Pentaquark photoproduction model

The differential cross section for the $\gamma p \to J/\psi\, p$ reaction can be written

$$\frac{d\sigma}{d\cos\theta} = \frac{4\pi\alpha}{32\pi s}\frac{p_f}{p_i}\frac{1}{4}\sum_{\lambda_\gamma,\lambda_p,\lambda_\psi,\lambda_{p'}} \left|\langle\lambda_\psi\lambda_{p'}\left|T\right|\lambda_\gamma\lambda_p\rangle\right|^2, \tag{1}$$

where p_i is the incoming momentum, p_f is the outgoing momentum, θ is the scattering angle, s is the standard Mandelstam variable, α is the electromagnetic fine structure constant, and λ_ψ, $\lambda_{p'}$, λ_γ and λ_p are the J/ψ, outgoing proton, photon and incoming proton helicities respectively. All the kinematical variables are defined in the center of mass reference system.

The amplitude $\langle\lambda_\psi\lambda_{p'}\left|T\right|\lambda_\gamma\lambda_p\rangle$ is constructed as the addition of two pieces: (i) the pentaquark resonance as a Breit–Wigner amplitude

$$\langle\lambda_\psi\lambda_{p'}|T_r|\lambda_\gamma\lambda_p\rangle = \frac{\langle\lambda_\psi\lambda_{p'}|T_{\rm dec}|\lambda_r\rangle\langle\lambda_r|T^\dagger_{\rm em}|\lambda_\gamma\lambda_p\rangle}{M_r^2 - W^2 - {\rm i}\Gamma_r M_r}, \tag{2}$$

where M_r, Γ_r and λ_r are the $P_c(4450)$ mass, width and helicity respectively, $W = \sqrt{s}$ is the invariant mass, $T_{\rm dec}$ stands for the strong vertex $pJ/\psi P_c$, and $T_{\rm em}$ stands for the electromagnetic vertex $\gamma p P_c$; and (ii) a nonresonant background described through an affective Pomeron exchange that is constrained by existing high-energy data [11]

$$\langle\lambda_\psi\lambda_{p'}|T_P|\lambda_\gamma\lambda_p\rangle = iA\left(\frac{s-s_t}{s_0}\right)^{\alpha(t)} e^{b_0(t-t_{\rm min})}\delta_{\lambda_p\lambda_{p'}}\delta_{\lambda_\psi\lambda_\gamma}, \tag{3}$$

where t is the standard Mandelstam variable, $t_{\rm min}$ is the minimum t for a given s, $s_0 = 1$ GeV^2 is the hadron physics scale, $\alpha(t) = \alpha_0 + \alpha' t$ is the Pomeron Regge trajectory whose parameters α_0 and α' are fitted to the data, and A, b_0 and t_0 are also fitted to the data. The electromagnetic vertex can be related to the helicity amplitudes $A_{1/2}$ and $A_{3/2}$ through

$$\langle\lambda_\gamma\lambda_p\left|T_{\rm em}\right|\lambda_r\rangle = \frac{W}{M_r}\sqrt{\frac{8M_N M_r \bar{p}_i}{4\pi\alpha}}\sqrt{\frac{\bar{p}_i}{p_i}}A_{1/2,3/2}, \tag{4}$$

where M_N is the nucleon mass and $\bar{p}_i$ is the momentum p_i evaluated at the resonance mass M_r. We do not know the helicity amplitudes of the pentaquark, hence, we need to make assumptions to describe this piece of the model. We assume the electromagnetic and strong vertices to be related by vector meson dominance

$$\langle\lambda_\gamma\lambda_p|T_{\rm em}|\lambda_r\rangle = \frac{\sqrt{4\pi\alpha}f_\psi}{M_\psi}\langle\lambda_\psi = \lambda_\gamma, \lambda_p|T_{\rm dec}|\lambda_r\rangle, \tag{5}$$

where M_ψ and $f_\psi = 280$ MeV are the J/ψ mass and decay constant respectively, and

$$\langle \lambda_\psi \lambda_{p'} | T_{\rm dec} | \lambda_r \rangle = g_{\lambda_\psi \lambda_{p'}} d^{J_r}_{\lambda_r, \lambda_\psi - \lambda_{p'}}(\cos\theta), \tag{6}$$

where $d^{J_r}_{\lambda_r, \lambda_\psi - \lambda_{p'}}(\cos\theta)$ are the Wigner d-functions with J_r the resonance spin, and $g_{\lambda_\psi, \lambda_{p'}}$ are the helicity couplings. We assume the three helicity couplings to be equal, $g = g_{\lambda_\psi, \lambda_{p'}}$. Hence the partial decay width $\Gamma_{\psi p}$ reads

$$\Gamma_{\psi p} = \mathcal{B}_{\psi p} \Gamma_r = \frac{\bar{p}_f}{32\pi^2 M_r^2} \frac{1}{2J_r + 1} \sum_{\lambda_R} \int d\Omega \, |\langle \lambda_\psi \lambda_{p'} | T_{dec} | \lambda_R \rangle|^2 = \frac{\bar{p}_f}{8\pi M_r^2} \frac{6g^2}{2J_r + 1}, \tag{7}$$

where $\mathcal{B}_{\psi p}$ is the $P_c(4450) \to J/\psi p$ branching ratio and $\bar{p}_f$ the p_f momentum at the resonance peak. Since there are no data in the resonance peak region we can only determine an upper limit for $\mathcal{B}_{\psi p}$ within a certain confidence level (CL) to ensure compatibility with the available data.

3. Results

We fit the available data on the $\gamma p \to J/\psi\, p$ reaction for $|t| \leq 1.5$ GeV^2 [11] and we compute the uncertainties in the parameters up to 68% and 95% CL using the bootstrap technique [12]. We also incorporate the uncertainty of the pentaquark mass and width through a Monte Carlo. In this way we can propagate exactly the experimental uncertainties to our predictions. Figure 1 shows the differential cross section $d\sigma/dt$ at $t = t_{\rm min}$ in the pentaquark energy region for the $J_r^P = 3/2^-$ and $5/2^+$ spin-parity assignments. The pentaquark peak is very pronounced, which implies that, if the LHCb signal is a true resonance and does not couple too weakly to the photon (as expected from vector meson dominance), it will be visible in the proposed photoproduction experiment at Jefferson Lab [9]. We found that data allow for the existence of a pentaquark whose $J/\psi p$ branching ratio has to be below 30% for $J^P = 3/2^-$ and below 17% for $J^P = 5/2^+$ at a 95% CL. The values of the parameters, their uncertainties, and the full extent of the analysis can be found in [10].

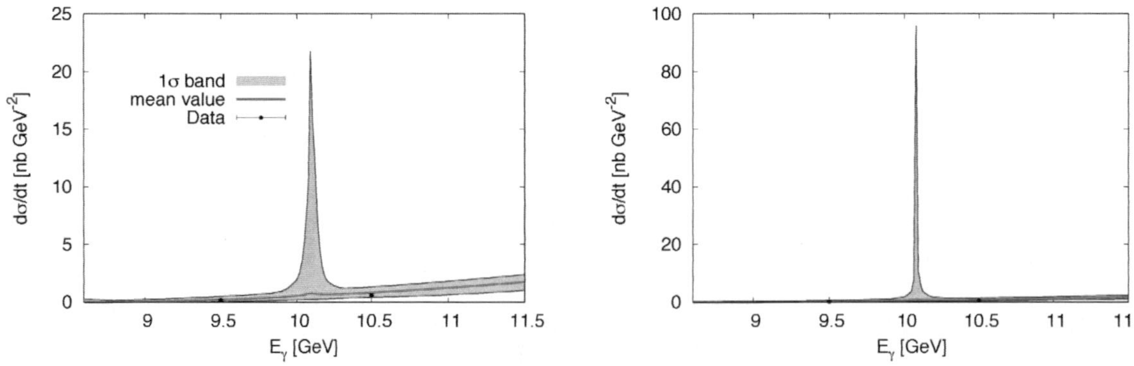

Figure 1. Differential cross section in the $P_c(4450)$ peak region at $t = t_{\rm min}$ for the $J_r = 3/2^-$ (left plot) and the $J_r = 5/2^+$ (right plot) spin-parity assignments.

4. Final remarks

The signals found in the $\Lambda_b^0 \to J/\psi K^- p$ decay by the LHCb collaboration [1] indicate the possible existence of a hidden charm pentaquark with a mass of 4450 MeV. One of the most promising ways to independently confirm the pentaquark nature of the signal consists on trying to photoproduce the state using the $\gamma p \to J/\psi p$ reaction. The recent approval by Jefferson Lab's PAC44 of this very experiment at Hall C [9] and the development of photoproduction models to analize the data [8, 10] make us believe that in the near future we will get a clear answer. The codes for the model in [10] can be downloaded and run online from the Joint Physics Analysis Center (JPAC) web page [13]. If the $P_c(4450)$ state is also detected in J/ψ photoproduction, it would likely be a compact state. Then, new experiments aimed to hunt the other members of this hidden-charm pentaquark multiplet can begin [14] as well as experiments to map the Q^2 dependence of the pentaquark electromagnetic form factor and provide insight into its internal structure.

Acknowledgments

This work is part of the efforts of the Joint Physics Analysis Center (JPAC) collaboration. CF-R work is supported in part by research grant IA101717 from PAPIIT-DGAPA (UNAM), by CONACYT (Mexico) grant No. 251817, and by Red Temática CONACYT de Física en Altas Energías (Red FAE, Mexico). ANHB research is supported in part by the Spanish Ministerio de Economía y Competitividad (MINECO), European FEDER funds under contract No. FIS2014-51948-C2-2-P, SEV-2014-0398, and Generalitat Valenciana under Contract PROMETEOII/2014/0068. AP research is supported by the U.S. Department of Energy, Office of Science, Office of Nuclear Physics under contract DE-AC05-06OR23177. CF-R thanks the organizers for their invitation to the conference and their warm hospitality at Cocoyoc.

References

[1] Aaij R *et al* (LHCb collaboration) 2015 *Phys. Rev. Lett.* **115** 072001
Aaij R *et al* (LHCb collaboration) 2016 *Phys. Rev. Lett.* **117** 082002
[2] Aaij R *et al* (LHCb collaboration) 2016 *Phys. Rev. Lett.* **117** 082003
[3] Patrignani C *et al* (Particle Data Group) 2016 *Chin. Phys.* C **40** 100001
[4] Fernández-Ramírez C *et al* 2016 *Phys. Rev.* D **93** 034029
[5] Meißner UG and Oller JA 2015 *Phys. Lett.* B **751** 59
Guo FK, Meißner UG, Wang W and Yang Z 2015 *Phys. Rev.* D **92** 071502
Guo FK, Meißner UG, Nieves J and Yang Z, 2016 *Eur. Phys. J.* A **52** 318
[6] Chen R, Liu X, Li Q and Zhu SL *Phys. Rev. Lett.* 115, 132002 (2015)
Chen HX *et al Phys. Rev. Lett.* 115, 172001 (2015)
Roca L, Nieves J and Oset E 2015 *Phys. Rev.* D **92** 094003
He J 2016 *Phys. Lett.* B **753** 547
Lü QF and Dong YB 2016 *Phys. Rev.* D **93** 074020
[7] Maiani L, Polosa AD and Riquer V 2015 *Phys. Lett.* B **749** 289
Lebed RF 2015 *Phys. Lett.* B **749** 454
Mironov A and Morozov A 2015 *JETP Lett.* **102** 271
Esposito A, Pilloni A and Polosa AD 2017 *Phys. Rep.* **668** 1
[8] Wang Q, Liu XH and Zhao Q 2015 *Phys. Rev.* D **92** 034022
Kubarovsky V and Voloshin MB 2015 *Phys. Rev.* D **92** 031502
Karliner M and Rosner JL 2016 *Phys. Lett.* B **752** 329
[9] Mezziani ZE *et al* (Hall C collaboration) 2016 *A Search for the LHCb Charmed "Pentaquark" using Photoproduction of J/ψ at Threshold in Hall C at Jefferson Lab*, *Preprint* 1609.00676 [hep-ex]
[10] Hiller Blin AN *et al* 2016 *Phys. Rev.* D **94** 034002
[11] Camerini U *et al* 1975 *Phys. Rev. Lett.* **35** 483
Ritson DM 1976 *AIP Conf. Proc.* **30** 75
Anderson RL 1976 *Madison Conf.* 102 *Preprint* SLAC-PUB-1741
Chekanov S *et al* (ZEUS collaboration) 2002 *Eur. Phys. J.* C **24** 345
Aktas A *et al* (H1 collaboration) 2006 *Eur. Phys. J.* C **46** 585
[12] Landay J *et al* 2017 *Phys. Rev.* C **95** 015203
[13] Mathieu V 2016 *AIP Conf. Proc.* **1735** 070004; `http://www.indiana.edu/~jpac/`
[14] Bijker R 2017 *J. Phys.: Conf. Ser.* submitted
Bijker R and Fernández-Ramírez C 2017 *Phys. Rev.* C submitted

IOP Conf. Series: Journal of Physics: Conf. Series **876** (2017) 012008 doi:10.1088/1742-6596/876/1/012008

Vibrations and potential energy surfaces (with Argonne V18) of ^{4}He and ^{3}He

Lorenzo Fortunato

Dipartimento di Fisica e Astronomia "G.Galilei" - Università di Padova
I.N.F.N. - Sez. di Padova,
via F.Marzolo 8, I-35131 Padova, Italy

E-mail: fortunat@pd.infn.it

Abstract. A potential energy surface is constructed for ^{3,4}He with the two-body Argonne V18 potential. The minimization suggests a semi-rigid asymmetric top structure for ^{4}He, where the appropriate point-group symmetry is C_2. We calculate the Hessian matrix, determining the 6 normal modes of vibration (in the range 300-700 MeV). The breathing mode is found to lie at too high an energy to be observable and the nature of the 0_2^+ excited states of the alpha particle at 20 MeV should probably be sought elsewhere. Similar investigations have been carried out for the A=3 system, finding a planar C_s configuration (scalene triangle) and three excited vibrational states (in the range 600-1900 MeV).

1. Introduction

In this contribution, the application of methods and techniques of quantum chemistry and molecular physics to the realm of nuclear physics is explored. This somewhat unconventional approach to nuclear structure has a tradition that goes back to early works in nuclear theory, such as those of Wheeler [1], Hafstad and Teller [2] and many others. Several studies on clusterization and exotic states of nuclear systems have relied on the same approach, for example the works of von Oertzen and collaborators [3, 4] where a Linear Combinations of Nucleonic Orbitals has been used, or works on the two center shell model [5]. Most of these approaches are nicely reviewed in Ref. [6] where details can be found about cluster models, Ikeda rules and several phenomenological approaches. Recently Bijker and Iachello have developed an algebraic version of cluster models that incorporates point-group symmetries and gives very interesting results on ^{12}C and ^{16}O [7, 8].

The common criticism to the application of molecular-type approaches to nuclei, revolves around the different stiffness of the two systems: most molecules are rigid (although there are examples of floppy molecules and of fluxional or super-rotor [9] molecules that baffle this traditional approach), while nuclei are 'liquid', in the sense that one expect large fluctuations inside the strongly interacting system, that forbid 'rigid' structures. Nevertheless, one normally adopts certain well-defined two-body interaction potentials, derived from phase-shifts analysis, between constituents of the nucleus that are deep enough to justify the calculation of a global potential energy surface (PES) built by using these interactions between all pairs of nucleons. This can be minimized giving equilibrium configurations that

IOP Conf. Series: Journal of Physics: Conf. Series **876** (2017) 012008 doi:10.1088/1742-6596/876/1/012008

have a precise geometric shape (or none). It is the equilibrium configurations, not the actual position of the nucleons, that displays symmetry properties that should be reflected in the symmetry of the wavefunctions and in the character of the vibrational states. We will analyze the cases of ^{4}He and ^{3}He (^{3}H is essentially equivalent), finding equilibrium shapes and their point-group symmetry and discussing the consequences on the vibrational spectrum. The energies of these modes will be found by applying the normal modes analysis (Hessian matrix) for the sake of simplicity, even though this is not the ideal approach because, despite the depths of the potential well being of the order of hundreds of MeV, the bound states are only a few MeV from the separation threshold, therefore they cannot be seen as deep lying states. It is instructive anyway to look at the results.

2. PES equilibrium configuration of ^{4}He

Let's consider the potential energy surface obtained by using the Argonne V18 potential between all the constituents of the α particle:

$$PES(\vec{r}_1, \vec{r}_2, \vec{r}_3, \vec{r}_4) = \sum_{i<j}^{4} V_{ij}(\mid \vec{r}_i - \vec{r}_j \mid) \tag{1}$$

where the vector coordinates $\vec{r}_i$ are not all independent in center of mass or intrinsic system, but they are constrained as follows:

$$\sum^{4} m_i \vec{r}_i = 0 \tag{2}$$

Of course the specific potentials are different for each type of interaction, i.e. they depend on the spin-isospin of each pair ij and on the orbital angular momentum. If we set 1 as the proton spin up, 2 as the proton spin down, 3 as the neutron spin up and 4 as the neutron spin down, the possible potential terms are listed in Table 1:

		spins	terms
1	2	p↑ p↓	${}^1S_0^{T=1}, {}^3P_J^{T=1}, {}^1D_2^{T=1}, \cdots$
1	3	p↑ n↑	${}^3S_1^{T=0}, {}^3P_J^{T=1}, {}^3D_J^{T=0}, \cdots$
1	4	p↑ n↓	${}^1S_0^{T=1}, {}^1P_0^{T=0}, {}^1D_2^{T=1}, \cdots$ ${}^3S_1^{T=0}, {}^3P_J^{T=1}, {}^3D_J^{T=0}, \cdots$
2	3	p↓ n↑	as in 1 4
2	4	p↓ n↓	as in 1 3
3	4	n↑ n↓	as in 1 2

Table 1. List of possible potential terms for each pair. Notice that the antisymmetry of the total wavefunction limits the possible combinations. The index J stands for all possible vector couplings between S and L.

The static picture that we adopt is that of looking for the minima of the potential energy surface with the combinations of potentials that minimize the energy. These are to be intended merely as equilibrium configurations for the nucleons, that nevertheless possess high momenta and have large fluctuations around these minima (as large as the nucleus itself). The dynamical picture instead is that of nucleons that are continuously scattered inside the α particle, and the scattering is described by the appropriate potential term. One might wonder if the PES corresponding to the lowest potentials (S-waves) ever cross

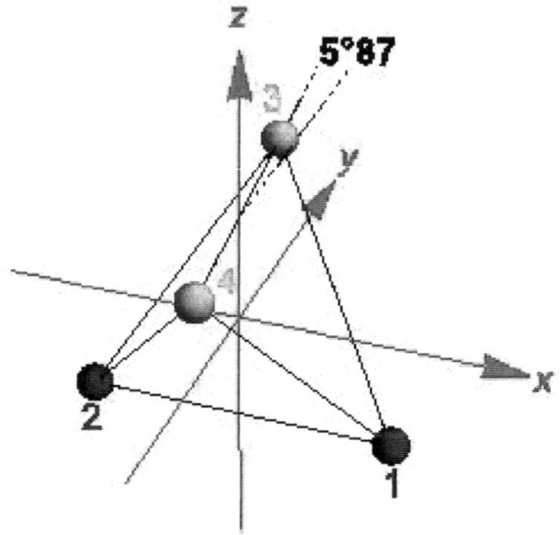

Figure 1. Equilibrium positions of protons (blue) and neutrons (green) in the α particle. Notice that the bond connecting 3 and 4, while perpendicular to the z-axis, is tilted 5^o87 away from the y-axis. This implies that the point-group symmetry is simply C_2.

(or has conical intersections) with higher order terms and this might lead to reaction paths, i.e. to loci of the potential energy surface that might allow a transition (chemists would call it isomerization) from a given geometric configuration (local minimum) to a different one (another local minimum) given a sufficient activation energy.

Let's use for the moment only the deepest possible S-wave potentials, i.e. those corresponding to the first configuration given in Tab. 1 for each pair. For the sake of simplicity I do not consider the ${}^3S_1^{T=0}$ configuration for the 1-4 and 2-3 pairs, because the depth of this potential is about one half of the ${}^1S_0^{T=1}$ one, therefore it does not affect the following search for the absolute minimum. Similarly, potentials with higher values of L have shallower pockets and can be disregarded for the present search. The PES depends in general on 4 vector positions, that correspond to 12 variables. A brute force minimization of Eq. (1) with constraints given by Eq. (2), gives an absolute minimum at -557.531 MeV. This minimum is spatially quite stiff with respect to small changes in each of the 12 variables and very easy to locate. It gives a somewhat simple geometric structure. A closer scrutiny of the coordinates that minimize the PES function reveals that a few simple rigid rotations can be made without altering this structure, nor the energy of the minimum. The following list of constraints can be imposed on minimization without loosing the minimum:

$$\begin{array}{cc} x_1 = -x_2 & x_3 = -x_4 \\ y_1 = -y_2 & y_3 = -y_4 \\ z_1 = z_2 & z_3 = z_4 \end{array}$$

The resulting geometric structure is depicted in Fig. 1: the equilibrium points of the two protons are contained in the xz plane and are parallel to the x axis, while those of the two neutrons lie in a line perpendicular to the z axis, but tilted 5.87^o away from the yz plane. The coordinates of the equilibrium points are:

$$\begin{array}{lll} x_1 = -0.437648 \text{ fm} & y_1 = 0 \text{ fm} & z_1 = -0.31883 \text{ fm} \\ x_3 = 0.0446687 \text{ fm} & y_3 = -0.43476 \text{ fm} & z_3 = 0.318391 \text{ fm} \end{array}$$

Note that one can be chosen to be zero without altering the minimization, in the c.m. frame it corresponds to rotate the whole system until the line connecting two particles lies in the $y = 0$ plane.

This shape is not a tetrahedron (it cannot be, the two protons repel each other more than the neutrons even neglecting spin), therefore T_d is excluded. If the inter-neutron bond would have been parallel to the y-axis a C_{2v} symmetry would have resulted, but this is not the case and the only possible symmetry operations are identity and a 180° rotation around the z-axis, therefore the point-group symmetry is simply C_2. The representations of this groups are very simple: A and B have positive and negative character with respect to the C_2 rotation. As far as we know, the only hint to non-tetrahedral shape of ^{4}He, was obtained in calculation of photonuclear sum rules using AV18 +UIX (notice that we don't have three-body potentials here) [10]. The authors conclude that helium-4 should look like a slightly deformed tetrahedron exactly as we find directly from the minimization.

1st Claim: the geometric shape of equilibrium positions of the nucleons composing the α particle, obtained by minimizing the the Potential Energy Surface (1) with the two-body AV18 potentials, is that of an asymmetric top with C_2 symmetry.

3. Vibrational spectrum in the harmonic approximation

We can perform a normal mode analysis in order to see the symmetry-type and energy of the fundamental vibrations. Even if the potentials are quite stiff, all of the following analysis is most certainly just an academic exercise that little has to do with reality: in fact, while in molecules masses distances and interactions often allow for the presence of tens of excited vibrational states, the nucleons in the α particle have such a different relationship between masses, distances and interactions that the energy of excited vibrational bands lies outside of the potential well. It is instructive anyway to take a look at this approach. If the potential energy is expanded in a Taylor series around the minimum, the first order derivatives vanish at the minimum and the mass-weighted Hessian matrix can be computed as

$$\tilde{H}_{lk} = H_{lk}/\sqrt{m_l m_k} = (\partial^2 PES/\partial q_l \partial q_k)/\sqrt{m_l m_k} \quad (3)$$

From group theory, after subtraction of translations and rotations, one should expect $3N-6$ normal modes, all non-degenerate: $\Gamma_{3N}-\Gamma_{rot}-\Gamma_{tra} = 4A\oplus 2B$. Of these, 4 are symmetric and 2 antisymmetric. We have obtained the explicit expressions for the displacements (eigenvectors of the Hessian) and energies (linked to the eigenvalues of the Hessian). The fundamental vibrations are all at very high energies, namely A 689.056 MeV, B 526.534 MeV, A 525.475 MeV, B 392.526 MeV, A 388.906 MeV, A 285.293 MeV, possibly too high to ever be observed as they are an order of magnitude higher than the various particle emission thresholds. They are depicted in Fig. 2. The first and higher in energy is the symmetric breathing mode. This is a very important result because in Ref. [11] the relevant and poignant question was asked whether the first excited 0_2^+ state of the alpha particle, lying at about 20 MeV might be interpreted as a breathing mode. Our analysis seems to exclude such a possibility, but it should be noted that a proper full calculation of the vibrational energy should be attempted before one can answer

IOP Conf. Series: Journal of Physics: Conf. Series **876** (2017) 012008 doi:10.1088/1742-6596/876/1/012008

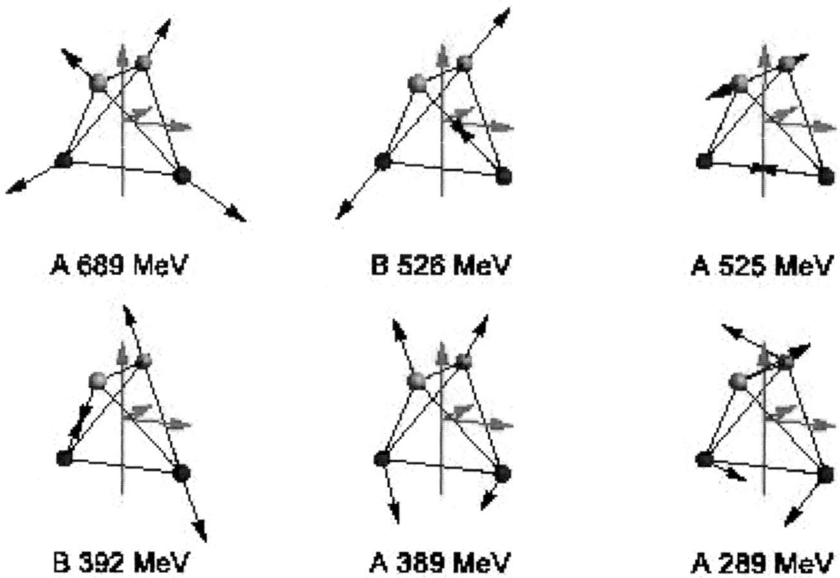

Figure 2. Normal modes of vibration of the C_2 configuration of the α particle.

this question with confidence. The two B modes at 526 and 392 MeV are both pn asymmetric stretches (one pn bond gets smaller, while the other gets stretched) that mix and repel, the A mode at 525 MeV is a pp asymmetric stretch (the pp bond gets smaller, while the other gets stretched and viceversa) and the two lowest A modes at 389 and 285 MeV are torsions that lift and simultaneously rotate the two neutrons away from the protons. The six sets of 4 displacement vectors form a system of normal coordinates, that could be taken as a symmetry-adapted basis for further calculations.

2st Claim: the fundamental vibrations of the C_2 configuration of the α particle, in the normal mode approximation, are too high in energy to be observed. Accordingly, the first excited 0_2^+ state at 20 MeV is probably not the breathing mode of this structure.

4. ^{3}He - C_s symmetry

Let's apply the same methods to helium-3. Analogous results can be obtained for the mirror system ^{3}H, that will not be reported here. The neutron and the two protons in the nucleus ^{3}He can be represented with three coordinates and three spin vectors: the two protons spin-up and down are assigned to 1 and 2 and the neutron (spin up or down is irrelevant) to 3. Thus the total PES can be written as:

$$PES(\vec{r}_1, \vec{r}_2, \vec{r}_3) = \sum_{i<j}^{3} V_{ij}(| \vec{r}_i - \vec{r}_j |) \tag{4}$$

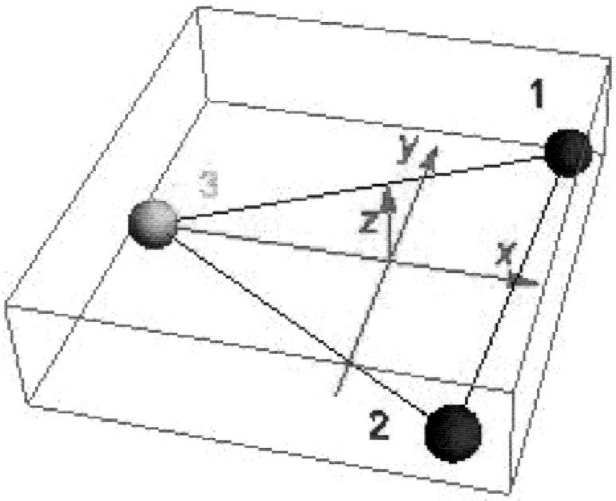

Figure 3. Equilibrium positions of 2 protons (blue) and a neutron (green) in ^{3}He. The point-group symmetry is simply C_s. Notice that the bond connecting 1 and 2 is not quite parallel to the y-axis, but tilted 3.31^o away from it.

The equilibrium configuration can be found by minimizing the above PES with the center of mass constraint, i.e.

$$\sum^{3} m_i \vec{r_i} = 0 \ . \tag{5}$$

The result of the minimization, with the minimum lying at -277.67 MeV, is not affected if we set some additional geometrical constraints, for example that all the three particles are in the $z = 0$ plane and that the neutron lies on the x-axis, namely :

$$y_1 = -y_2$$
$$z_1 = z_2 = z_3 = 0$$

The resulting geometric structure is depicted in Fig. 3: the triangle is not equilateral, as one could have expected from the fact that one of the particle is different from the other two, but surprisingly is not even isosceles, all sides are dissimilar. This is due to the fact that the interactions of the n with the p's is slightly different depending on the relative spin orientations. The coordinates of the equilibrium points are:

$$\begin{array}{lll} x_1 = \ 0.283102 \text{ fm} & y_1 = \ 0.436916 \text{ fm} & z_1 = 0. \text{ fm} \\ x_2 = \ 0.232502 \text{ fm} & y_2 = -0.436916 \text{ fm} & z_2 = 0. \text{ fm} \\ x_3 = -0.514895 \text{ fm} & y_3 = 0. \text{ fm} & z_3 = 0. \text{ fm} \end{array}$$

The point-group symmetry is just C_s, i.e. the group of identity and reflection upon the horizontal plane, with the possible representations A' and A'' that differ because of the sign under reflection.

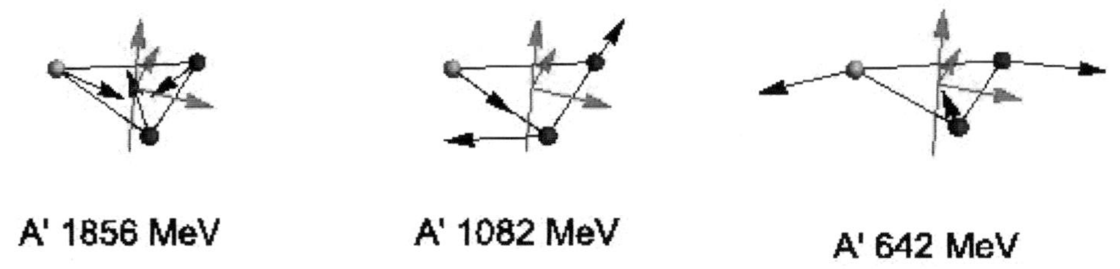

Figure 4. Normal modes of vibration of the C_2 configuration of the ^{3}He nucleus. Notice that all displacement vectors have unequal lengths and that their orientations are not interchangeable by any rotation.

The group theoretical analysis of the normal modes after subtraction of translations and rotations, gives $3N - 6$ normal modes: $\Gamma_{3N} - \Gamma_{rot} - \Gamma_{tra} = 3A'$, all symmetric, therefore all preserving the overall scalene shape.

3rd Claim: the geometric shape of equilibrium positions of the nucleons composing ^{3}He, obtained by minimizing the the Potential Energy Surface (4) with the two-body AV18 potentials, is that of a scalene triangle with C_s symmetry.

The energies of the fundamental vibrations obtained from diagonalization of the Hessian matrix are: 642.033, 1081.92 and 1856.07 MeV. The three sets of displacement vectors shown in Fig. 4 indicate that the lowest vibration is a sort of bending mode, the intermediate one is a sort of stretching mode and the higher one is a sort of breathing mode. We have slavishly repeated the words 'sort of' to underline that, in general, all nucleons have displacements of different lengths, for example the two protons in the breathing mode don't move in the same way or to the same extent. The above energies of the excited vibrational states are very high for low-energy nuclear structure, far beyond proton or neutron separation threshold and pion production threshold. Probably one could wonder if the harmonic approximation makes any sense, but I expect that the normal modes analysis gives results that are at least of the same order of magnitude as the full ab initio calculation of eigen-energies. These states could exist as resonances, but their width would be extremely large and therefore I do not expect that they could ever be observed in scattering or photoabsorption experiments.

5. Conclusions

The exact ab initio calculations of the vibrational energies would be a major step forward in this approach, because it might tell (by exclusion for instance) if any of the observed excited states of the α particle has anything to do with a breathing mode. Extensions of the present approach to heavier systems are in principle not difficult, provided that one considers also higher order interaction terms that go beyond $L = 0$. It would be very nice to explore if the PES with at least one D interaction term give energy differences with the ground state PES that are in the range of 20 MeV and if there is any conical intersection, i.e. if two different geometrical equilibrium shapes have PES that touch or cross each other (they have a locus of degenerate energies) where interesting phenomena might appear.

IOP Conf. Series: Journal of Physics: Conf. Series **876** (2017) 012008 doi:10.1088/1742-6596/876/1/012008

References

[1] Wheeler J A 1937 *Phys. Rev.* **52** 1083-1106
[2] Hafstad L R and Teller E 1938 *Phys.Rev.* **54** 681
[3] von Oertzen W 1970 *Nucl. Phys. A* **148** 529
[4] von Oerzen W and Bohlen H G 1975 *Phys. Rep. C* **19** 1
[5] Freer M and Merchant A C 1997 *J. Phys. G* **23** 261
[6] von Oertzen W, Freer M and Kanada-En'yo Y 2006 *Phys. Rep.* **432** 43-113
[7] Bijker R and Iachello F 2014 *Phys. Rev. Lett.* **112** 152501
[8] Bijker R and Iachello F 2017 *Nucl. Phys. A* **957** 154176
[9] Schmiedt H, Jensen P and Schlemmer S 2017 *Chem. Phys. Lett.* **672** 34-46
[10] Gazit D, Barnea N, Bacca S, Leidemann W and Orlandini G 2006 *Phys. Rev. C* **74** 061001(R)
[11] Bacca S, Barnea N, Leidemann W and Orlandini G 2105 *Phys. Rev. C* **91** 024303

Current and Future Tests of the Algebraic Cluster Model of ^{12}C

Moshe Gai
LNS at Avery Point, University of Connecticut, 1084 Shennecossett Rd., Groton, CT 06340.
E-mail: moshe.gai@uconn.edu

Abstract. A new theoretical approach to clustering in the frame of the Algebraic Cluster Model (ACM) has been developed. It predicts, in ^{12}C, rotation-vibration structure with rotational bands of an oblate equilateral triangular symmetric spinning top with a $\mathcal{D}_{3h}$ symmetry characterized by the sequence of states: $0^+, 2^+, 3^-, 4^{\pm}, 5^-$ with a degenerate 4^+ and 4^- (parity doublet) states. Our newly measured 2^+_2 state in ^{12}C allows the first study of rotation-vibration structure in ^{12}C. The newly measured 5^- state and 4^- states fit very well the predicted ground state rotational band structure with the predicted sequence of states: $0^+, 2^+, 3^-, 4^{\pm}, 5^-$ with almost degenerate 4^+ and 4^- (parity doublet) states. Such a $\mathcal{D}_{3h}$ symmetry is characteristic of triatomic molecules, but it is observed in the ground state rotational band of ^{12}C for the first time in a nucleus. We discuss predictions of the ACM of other rotation-vibration bands in ^{12}C such as the (0^+) Hoyle band and the (1^-) bending mode with prediction of ("missing 3^- and 4^-") states that may shed new light on clustering in ^{12}C and light nuclei. In particular, the observation (or non observation) of the predicted ("missing") states in the Hoyle band will allow us to conclude the geometrical arrangement of the three alpha particles composing the Hoyle state at 7.6542 MeV in ^{12}C. We discuss proposed research programs at the Darmstadt S-DALINAC and at the newly constructed ELI-NP facility near Bucharest to test the predictions of the ACM in isotopes of carbon.

1. Introduction

A recent experiment performed at the HIγS gamma-ray facility [1] using the optical readout TPC (O-TPC) [2] provided unambiguous identification of the 2^+ member of the Hoyle rotational band [3] as shown in Fig.1. This observation allows, for the first time, the study of the predicted rotation-vibration structure of ^{12}C.

The structure of light nuclei, and specifically of clustering received new interest with major developments of ab-initio calculations of light nuclei and in particular of ^{12}C. Theoretical ab-initio shell-model [4] and symmetry inspired shell-model [5] calculations as well as ab-intio Effective Field Theory (EFT) calculations on the lattice [6, 7], Fermionic Molecular Dynamics (FMD) model [8] and Anti-symmetrized Molecular Dynamics (AMD) model [9], are employed to yield a microscopic foundation of the clustering phenomena that naturally occur in cluster models [10, 11]. For example, one issue of current concern [12] is the geometrical structure of the three alpha-particles in the Hoyle state at 7.6542 MeV in ^{12}C (linear chain, obtuse triangle or equilateral triangle) and the Hoyle rotational band built on top of the Hoyle state [13, 14, 15, 16, 17].

Cluster states are best described as molecular states which are characterized by the separation (Jacobi) vector(s) connecting the constituent objects. For diatomic like object one single

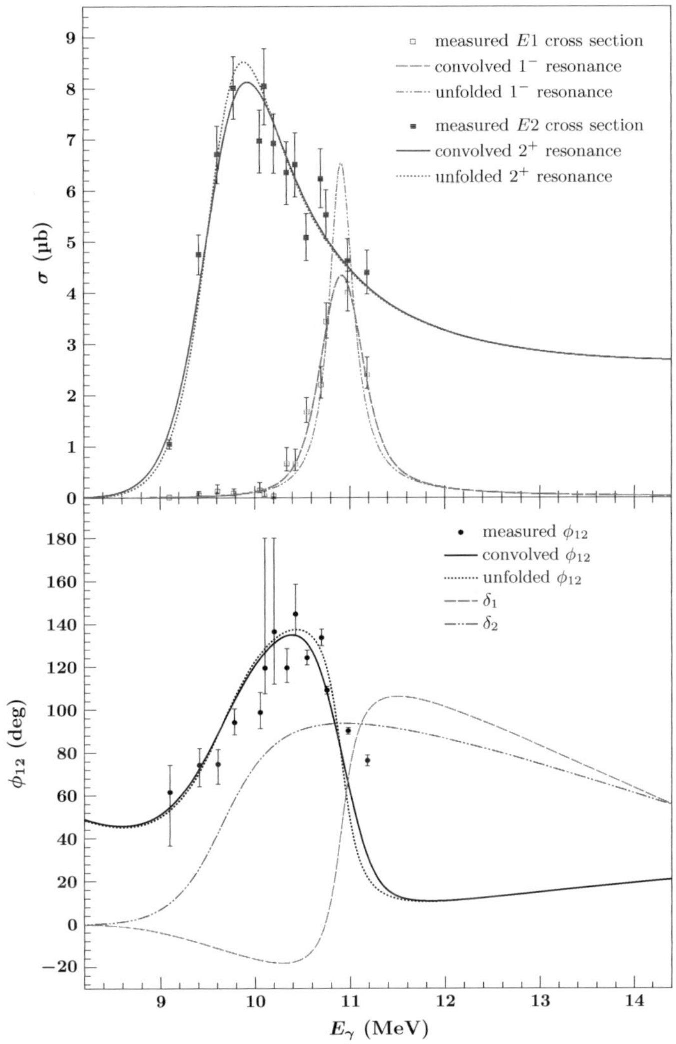

Figure 1. The measured cross section of the $^{12}C(\gamma, \alpha_0)^8Be$ reaction [3] and the E1-E2 phase angle, ϕ_{12}, with a fit including the E1 and E2 components.

(separation) vector is required leading to the predicted U(4) symmetry [18] that was observed in ^{18}O via the characteristic enhanced E1 decays [19]. Triatomic symmetric spinning tops are characterized by the two perpendicular Jacobi vectors (λ and ρ) leading to the predicted U(7) symmetry with the geometrical $\mathcal{D}_{3h}$ symmetry [20, 21].

The observation of the 2^+ Hoyle rotational excitation in ^{12}C [3, 14] together with the recently discovered 4^- [22, 23] and 4^+ [24] states in ^{12}C are in agreement with the predicted spectrum of an oblate spinning top with a $\mathcal{D}_{3h}$ symmetry [20, 21]. It was predicted [20, 21] that the three alpha-particle system of ^{12}C leads to the ground state rotational band including the most unusual sequence of states: $0^+, 2^+, 3^-, 4^\pm, 5^-$. The new high spin 5^- state [25] as well the previously published 4^- state [22, 23] lead to a J(J+1) trajectory as predicted by this U(7) model [25] including the nearly degenerate 4^- and 4^+ states as shown in Fig. 2.

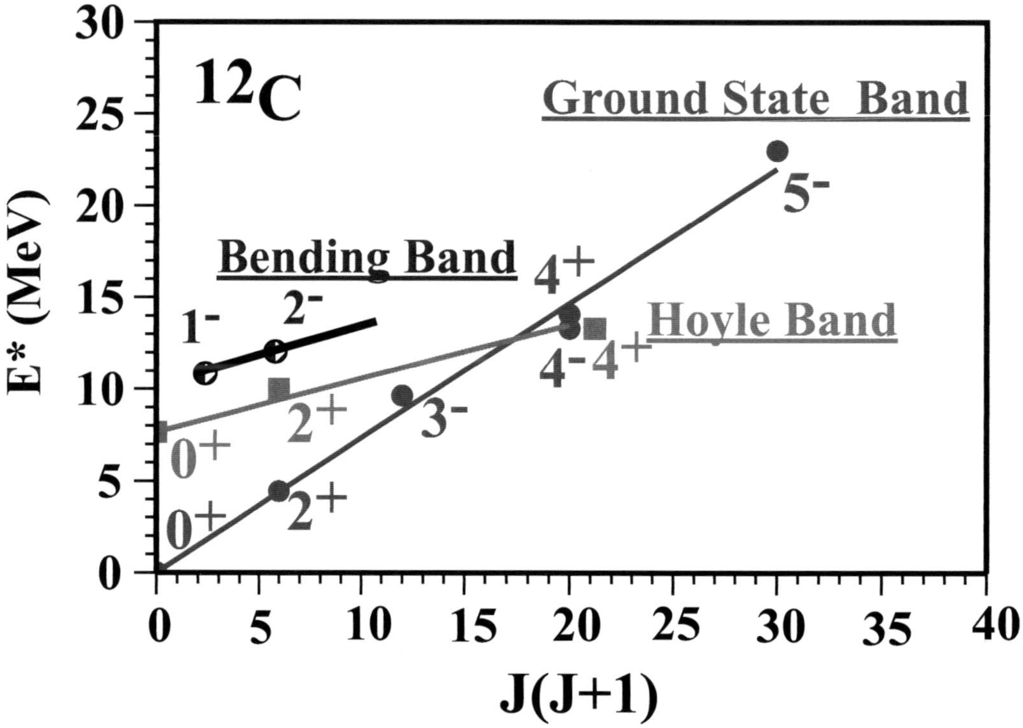

Figure 2. Rotational band structure of the: ground-state band, the Hoyle band and the bending vibration band in ^{12}C [25].

2. The Algebraic Cluster Model

The spectrum of ^{12}C predicted by the Algebraic Cluster Model (ACM) [20, 21] is shown in Fig. 3 where it is also compared to the measured spectrum of ^{12}C [25]. In addition to the ground-state rotational band this U(7) model [20, 21] predicts the Hoyle state at 7.6542 MeV in ^{12}C to be the first vibrational breathing mode of the three alpha-particle equilateral configuration leading to the same rotational structure albeit with a larger moment of inertia (by a factor of 2). Recent measurements revealed the 2^+ [14] and 4^+ [24] members of the Hoyle rotational band and we are currently searching [26, 27] for the 4^- predicted by the ACM to be nearly degenerated with the 4^+ state and the 3^- (broad) state that was suggested to lie between 11 and 14 MeV [22, 28]. The observation (or lack there) of these "missing" states will allow us to determine whether the Hoyle state is composed of three alpha-particles in an equilateral triangle arrangement [25, 8, 9] or an obtuse triangle [7] or whether it is better described as a vibrational excitation of a "diffuse gas" of three alpha-particles [11].

The U(7) model also predicts the 1^- state at 10.84 MeV to be the vibrational bending mode with a rotational band including the 1^- anddegenerate $2^{\pm}$ states. We are searching [26, 27] for the third 2^+ of ^{12}C that is predicted by the U(7) model to lie near the observed 2^- state at 11.8 MeV.

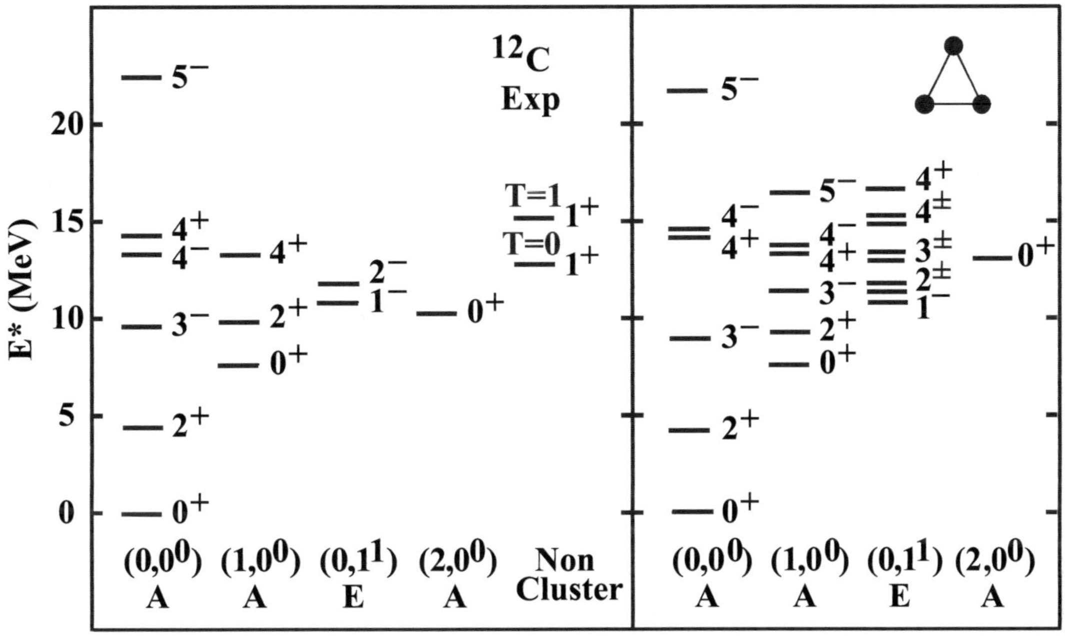

Figure 3. Comparison between the low-lying experimental spectrum of ^{12}C [25] and the energies of the oblate symmetric top. The levels are organized in columns corresponding to the ground state band and the vibrational bands with A and E symmetry [20, 21] of an oblate top with triangular symmetry. The last column on the left-hand side, shows the lowest observed non-cluster (1^+) levels.

3. Future Tests of the ACM

We propose a new research program in Nuclear Structure studies to be performed at the ELI-NP facility near Bucharest, Romania, with a newly proposed electronic readout TPC (eTPC) and a Silicon Strip Detector (SSD) [27]. It is proposed to measure of the multi-alpha decay of ^{12}C and ^{16}O with a TPC detector. Such measurements of multi-alpha decay of ^{12}C were also performed in the past with Silicon Strip Detectors (SSD) [22, 23, 24], hence, similar measurements with SSD are also considered for the ELI-NP. We emphasize that the measurements discussed for ^{12}C and ^{16}O are considered as typical examples of studies in nuclear structure that can be performed with our proposed detectors at the ELI-NP and such measurements in other light nuclei will lead to new perspectives on clustering phenomena in light nuclei.

4. Conclusions

In conclusion the ACM appears to open a new chapter in cluster physics of light nuclei and it presents an opportunity for further experimental investigation of light nuclei. The reader is referred to the Technical Design Report (TDR) [27] for a discussion of the scientific goals of the Charged Particle Working Group (CPWG) of the ELI-NP facility in the studies of light nuclei.

5. Acknwledgement

The material presented in this paper is based upon work supported by the U.S. Department of Energy, Office of Science, Office of Nuclear Physics, under Award Number DE-FG02-94ER40870.

References

[1] H. R. Weller, M. W. Ahmed, H. Gao, W. Tornow, Y. K. Wu, M. Gai, and R. Miskimen, Prog. Part. Nucl. Phys. **62**, 257 (2009).
[2] M. Gai, M.W. Ahmed, S.C. Stave, W.R. Zimmerman, A. Breskin, B. Bromberger, R. Chechik, V. Dangendorf, Th. Delbar, R.H. France III, S.S. Henshaw, T.J. Kading, P.P. Martel, J.E.R. McDonald, P.-N. Seo, K. Tittelmeier, H.R. Weller and A.H. Young; JINST **5**, 12004 (2010).
[3] W.R. Zimmerman, M.W. Ahmed, B. Bromberger, S.C. Stave, A. Breskin, V. Dangendorf, Th. Delbar, M. Gai, S.S. Henshaw, J.M. Mueller, C. Sun, K. Tittelmeier, H.R. Weller, and Y.K. Wu; Phys. Rev. Lett. **110**, 152502 (2013).
[4] R. Roth, J. Langhammer, A. Calci, S. Binder, and P. Navratil; Phys. Rev. Lett. **107**, 072501 (2011).
[5] A. C. Dreyfuss, K. D. Launey, T. Dytrych, J. P. Draayer, and C. Bahri, Phys. Lett. B **727**, 511 (2013).
[6] E. Epelbaum, H. Krebs, D. Lee, and U.-G. Meissner, Phys. Rev. Lett. **106**, 192501 (2011).
[7] E. Epelbaum, H. Krebs, T. A. Lahde, D. Lee, and U.-G. Meissner, Phys. Rev. Lett. **109**, 252501 (2012).
[8] M. Chernykh, H. Feldmeier, T. Neff, P. von Neumann-Cosel, and A. Richter, Phys. Rev. Lett. **98**, 032501 (2007).
[9] Yoshiko Kanada-Enyo, Prog. of Theo. Phys. **117**, 655 (2007).
[10] M. Kamimura, Nucl. Phys. **A351**, 456 (1981).
[11] Y. Funaki, H. Horiuchi, W. von Oertzen, G. Ropke, P. Schuck, A. Tohsaki, and T. Yamada, Phys. Rev. C **80**, 064326 (2009).
[12] H.O.U. Fynbo and M. Freer, Physics **4**, 94 (2011).
[13] H. Morinaga, Phys. Rev. **101**, 254 (1956).
[14] M. Itoh et al., Nucl. Phys. **A738**, 268 (2004); Phys. Rev. C **84**, 054308 (2011).
[15] M. Freer et al., Phys. Rev. C **80**, 041303(R) (2009).
[16] W. R. Zimmerman, N. E. Destefano, M. Freer, M. Gai, and F. D. Smit, Phys. Rev. C **84**, 027304 (2011).
[17] M. Freer et al., Phys. Rev. C **86**, 034320 (2012).
[18] F. Iachello; Phys. Rev. C **23**, 2778 (1981).
[19] M. Gai, M. Ruscev, A.C. Hayes, J.F. Ennis, R. Keddy, E.C. Schloemer, S.M. Sterbenz and D.A. Bromley; Phys. Rev. Lett. **50**, 239 (1983).
[20] R. Bijker and F. Iachello; Phys. Rev. C **61**, 067305 (2000).
[21] R. Bijker and F. Iachello; Ann. Phys. **298**, 334 (2002).
[22] M. Freer *et al.*; Phys. Rev. C **76**, 034320 (2007).
[23] O.S. Kirsebom *et al.*; Phys. Rev. C **81**, 064313 (2010).
[24] M. Freer *et al.*; Phys. Rev. C **83**, 034314 (2011).
[25] D.J. Marin-Lambarri, R. Bijker, M. Freer, M. Gai, Tz. Kokalova, D.J. Parker, C. Wheldon; Phys. Rev. Lett. **113**, 012502 (2014).
[26] A 12C(e,e') experiment in progress at Darmstadt, P. von Neumann-Cosel, private communication, 2013.
[27] D. Filipescu, A. Anzalone, D.L. Balabanski, S.S. Belyshev, F. Camera, M. La Cognata, P. Constantin, L. Csige, P.V. Cuong, M. Cwiok, V. Derya, W. Dominik, M. Gai, S. Gales, I. Gheorghe, B.S. Ishkhanov, A. Krasznahorkay, A.A. Kuznetsov, C. Mazzocchi, V.N. Orlin, N. Pietralla, M. Sin, C. Spitaleri, K.A. Stopani, O. Tesileanu, C.A. Ur, I. Ursu, H. Utsunomiya, V.V. Varlamov, H.R. Weller, N.V. Zamfir, and A. Zilges; Eur. Phys. J. A**51**, 185 (2015).
[28] O.S. Kirsebom; CHANDA, workshop on target preparations, November 23 25, 2015, Paul Scherrer Institute, Viligen, Switzerland.

IOP Conf. Series: Journal of Physics: Conf. Series **876** (2017) 012010 doi:10.1088/1742-6596/876/1/012010

The baryon spectroscopy: strong decays and strange suppression[1]

H. García-Tecocoatzi

INFN sezione di Genova and Universitá di Genova, via Dodecaneso 33, I-16146 Genova, Italy.

E-mail: hgarcia@ge.infn.it

Abstract. In this contribution, we present the open-flavor strong decays of light baryons computed within the framework of quark model. The transition amplitudes are computed using a modified 3_0^P operator, where a mechanism strange suppression is taken into account. Also we discus the strange suppression within an extension of the quark model.

1. Introduction

At the moment, the number of known light-quark mesons is much larger than the number of known baryon resonances [1]. However, it is known that the baryon spectrum is much more complex than the meson one. For instance, it is difficult to identify those high-lying baryon resonances that are only weakly coupled to the $N\pi$ channel [2, 3], since they cannot be seen in elastic $N\pi$ scattering experiments. Regarding strong decays of baryons no satisfactory description has yet been achieved. We could list several problems, for instance, the QCD mechanism behind the OZI-allowed strong decays [4] is still not clear. Theoretical calculations of baryon strong, electromagnetic and weak decays can still help the experimentalists in their search of those resonances that are still unknown.

In this contribution, we first discuss a strange suppression mechanism in the open-flavor strong decays of light baryons within the quark model framework. The quark model (QM) [8, 9, 10, 11, 12, 13, 14, 15, 16, 17] can reproduce the behavior of observables such as the spectrum and the magnetic moments in the baryon and meson sector. The decay widths of baryon resonances into baryon-pseudoscalar meson pairs were recently reported in Ref. [5], within the $^3P_0$ decay model framework [6, 7], using the mass spectrum of two different models: the $U(7)$ algebraic model [17], by Bijker, Iachello and Leviatan, and the hypercentral model (hQM) [12], developed by Giannini and Santopinto.

Finally, we discuss one of the latest applications of the Unquenchend Quark Model (this approach is a generalization of the unitarized quark model [18, 19, 20, 21, 22]) to describe the strangeness suppression in the electro-production of resonances from proton [23] within the UQM framework. The unquenching of the quark model for hadrons is a way to take into account the continuum-coupling effects. Above threshold, these couplings lead to strong decays and below threshold, they lead to virtual $q\bar{q} - q\bar{q}$ ($qqq - q\bar{q}$) components in the hadron wave function and shifts of the physical mass with respect to the bare mass.

[1] Invited talk presented at Symposium on Nuclear Physics, January 4-7 2017, Cocoyoc(Mexico).

IOP Conf. Series: Journal of Physics: Conf. Series **876** (2017) 012010 doi:10.1088/1742-6596/876/1/012010

2. The 3_0^P decay model with strangeness suppresion

Here, we present the formalism to compute the two-body strong decay widths of baryons resonances in the $^3P_0$ pair-creation model, when a strangeness suppression mechanism is included. The decay widths are computed as [5, 6, 7, 24, 25, 26]

$$\Gamma_{A\to BC} = \Phi_{A\to BC}(q_0) \sum_{\ell,J} \left| \langle BC\vec{q}_0\,\ell J|\,T^\dagger\,|A\rangle \right|^2 , \tag{1}$$

where, $\Phi_{A\to BC}(q_0)$ is the relativistic phase space factor:

$$\Phi_{A\to BC}(q_0) = 2\pi q_0 \frac{E_b(q_0)E_c(q_0)}{M_a} , \tag{2}$$

depending on q_0 and on the energies of the two intermediate state hadrons, $E_b = \sqrt{M_b^2 + q_0^2}$ and $E_c = \sqrt{M_c^2 + q_0^2}$. We assumed harmonic oscillator wave functions, depending on a single oscillator parameter $\alpha_{\rm b}$ for the baryons and $\alpha_{\rm m}$ for the mesons. The coupling between the final state hadrons $|B\rangle$ and $|C\rangle$ is described in terms of a spherical basis [5]. Specifically, the final state $|BC\vec{q}_0\,\ell J\rangle$ can be written as

$$\begin{aligned} |BC\vec{q}_0\,\ell J\rangle &= \sum_{m,M_b,M_c} \langle J_bM_bJ_cM_c|\,J_{bc}M_{bc}\rangle\,\langle J_{bc}M_{bc}\ell m\,|JM\rangle\,\frac{Y_{\ell m}(\hat{q})}{q^2}\delta(q-q_0) \\ &\times\ |(S_b,L_b)J_bM_b\rangle\,|(S_c,L_c)J_cM_c\rangle\,, \end{aligned} \tag{3}$$

where the ket $|BC\vec{q}_0\,\ell J\rangle$ is characterized by a relative orbital angular momentum ℓ between B and C and a total angular momentum $\vec{J} = \vec{J}_b + \vec{J}_c + \vec{\ell}$.

The transition operator of the $^3P_0$ model is given by [5, 24, 25, 26]:

$$\begin{aligned} T^\dagger &= -3\,\gamma_0^{\rm eff} \int d\vec{p}_4\,d\vec{p}_5\,\delta(\vec{p}_4+\vec{p}_5)\,C_{45}\,F_{45}\,e^{-r_q^2(\vec{p}_4-\vec{p}_5)^2/6} \\ &\quad [\chi_{45}\times\mathcal{Y}_1(\vec{p}_4-\vec{p}_5)]_0^{(0)}\,b_4^\dagger(\vec{p}_4)\,d_5^\dagger(\vec{p}_5)\ . \end{aligned} \tag{4}$$

Here, $b_4^\dagger(\vec{p}_4)$ and $d_5^\dagger(\vec{p}_5)$ are the creation operators for a quark and an antiquark with momenta $\vec{p}_4$ and $\vec{p}_5$, respectively. The $q\bar{q}$ pair is characterized by a color singlet wave function C_{45}, a flavor singlet wave function F_{45}, a spin triplet wave function χ_{45} with spin $S=1$ and a solid spherical harmonic $\mathcal{Y}_1(\vec{p}_4-\vec{p}_5)$, since the quark and antiquark are in a relative P wave. The operator $\gamma_0^{\rm eff}$ of Eq. (4) is the effective pair-creation strength $\gamma_0^{\rm eff}$ [5, 24, 25, 26, 28], defined as

$$\gamma_0^{\rm eff} = \frac{m_n}{m_i}\,\gamma_0, \tag{5}$$

is introduced, with $i = n$ (i.e. u or d) or s. In our recent study [5], we performed the correct treatment of $\gamma_0^{\rm eff}$ in the open flavor strong decays. We showed that $\gamma_0^{\rm eff}$can be absorbed in the flavor couplings, thus the flavor singlet wave function is change as follow

$$\begin{aligned} \gamma_0^{\rm eff}\phi_0 &= \gamma_0^{\rm eff}\frac{1}{\sqrt{3}}\left[|u\bar{u}\rangle + |d\bar{d}\rangle + |s\bar{s}\rangle\right] \\ &\to\ \gamma_0\phi_0^{\rm eff} = \gamma_0\frac{|u\bar{u}\rangle+|d\bar{d}\rangle+\frac{m_n}{m_s}|s\bar{s}\rangle}{\sqrt{2+\left(\frac{m_n}{m_s}\right)^2}}\ . \end{aligned} \tag{6}$$

IOP Conf. Series: Journal of Physics: Conf. Series **876** (2017) 012010 doi:10.1088/1742-6596/876/1/012010

3. UQM

In the unquenched quark model for baryons [30, 31, 32, 33] and mesons [25, 24, 26, 27], the hadron wave function is made up of a zeroth order qqq ($q\bar{q}$) configuration plus a sum over the possible higher Fock components, due to the creation of $^3P_0$ $q\bar{q}$ pairs. Thus, we have

$$\begin{aligned} \mid \psi_A \rangle &= \mathcal{N} \left[\mid A \rangle + \sum_{BC\ell J} \int d\vec{K}\, k^2 dk \mid BC\ell J; \vec{K} k \rangle \right. \\ & \left. \qquad \frac{\langle BC\ell J; \vec{K} k \mid T^\dagger \mid A \rangle}{E_a - E_b - E_c} \right] , \end{aligned} \tag{7}$$

where $T^\dagger$ stands for the $^3P_0$ quark-antiquark pair-creation operator [24, 25, 26, 27], A is the baryon/meson, B and C represent the intermediate state hadrons, see Figures 1 and 2. E_a, E_b and E_c are the corresponding energies, k and ℓ the relative radial momentum and orbital angular momentum between B and C and $\vec{J} = \vec{J}_b + \vec{J}_c + \vec{\ell}$ is the total angular momentum. It is worthwhile noting that in Refs. [24, 25, 26, 27, 28], the constant pair-creation strength in the operator (7) was substituted with an effective one, to suppress unphysical heavy quark pair-creation.

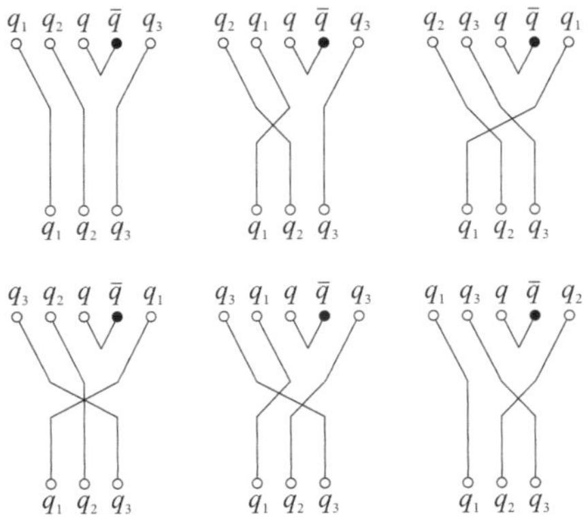

Figure 1. Quark line diagrams for $A \to BC$ with $q\bar{q} = s\bar{s}$ and $q_1q_2q_3 = uud$

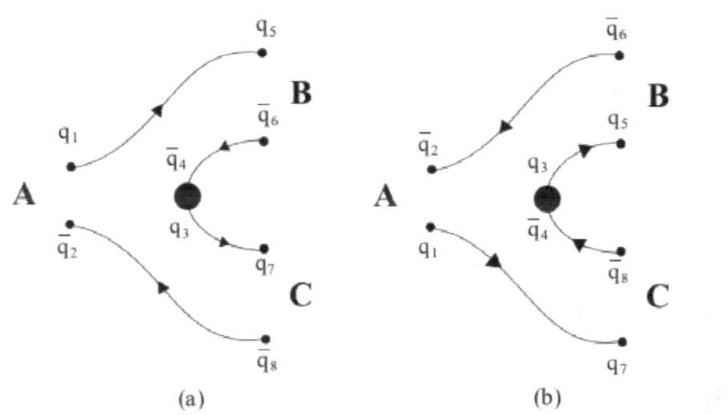

Figure 2. Two diagrams can contribute to the process $A \to BC$. q_i and q_i stand for the various initial (i = 1 - 4) and final (i = 5 - 8) quarks or antiquarks, respectively.

The introduction of coupling continuum effects in the QM has been essential to study observables that only depend on $q\bar{q}$ sea pairs, like the strangeness content of the nucleon electromagnetic form factors [29, 30] or the flavor asymmetry of the nucleon sea [31]. In other cases, continuum effects can provide important corrections to baryon/meson observables, like the self-energy corrections to meson masses [24, 25, 26, 27] or the importance of the orbital angular momentum in the spin of the proton [32].

4. Strengeness suppression in the electro-production

The UQM wave function can be tested in the production ratios of baryon-meson states [23]. In Ref. [23] was shown that the production rates can be expressed as the product of a spin-flavor-isospin factor and a radial integral

$$\frac{p \to \Lambda K^+}{p \to n\pi^+} = \frac{27}{50} \frac{I_{N\to\Lambda K}}{I_{N\to N\pi}} , \tag{8}$$

Table 1. Comparison of the strong decay widths with/without strangeness suppression mechanism.

Decay mode	Model	Without suppression	With suppresion [5]	Exp [1]
$N(1710) \to \Lambda K$	U(7)	8	3	3-63
	hQM	39	14	
$N(1720) \to \Lambda K$	U(7)	39	14	2-60
	hQM	33	12	
$N(1900) \to \Lambda K$	U(7)	36	13	0-37
	hQM	36	13	
$N(1900) \to \Sigma K$	U(7)	3	1	6-26
	hQM	3	1	
$\Delta(1910) \to \Sigma K$	U(7)	105	38	9-48
	hQM	105	38	
$\Delta(1920) \to \Sigma K$	U(7)	64	23	3-7
	hQM	61	22	
$\Delta(1950) \to \Sigma K$	U(7)	14	4	1-2
	hQM	8	3	
$\Sigma^*(2030) \to \Xi K$	U(7)	208	75	26-46

Table 2. Ratios of electro-production cross sections.

Ratio	UQM [23]	Exp. [34]
$p \to \Lambda K^+/p \to n\pi^+$	0.227	$0.19 \pm 0.01 \pm 0.03$
$p \to \Lambda K^+/p \to p\pi^0$	0.454	$0.50 \pm 0.02 \pm 0.12$
$p \to p\pi^0/p \to n\pi^+$	0.500	$0.43 \pm 0.01 \pm 0.09$

with

$$I_{A\to BC} = \int_0^\infty \frac{k^4 e^{-2F^2k^2}}{\Delta E^2_{A\to BC}(k)} dk \ . \tag{9}$$

Here, the energy denominator represents the energy difference between initial and final hadrons calculated in the rest frame of the initial baryon A. The value of F^2 depends on the size of the harmonic oscillator wave functions for baryons and mesons, and the Gaussian smearing of the pair-creation vertex, and its value is taken from [30] to be $F^2 = 2.275$ GeV^{-2}.

5. Results and discussion

The strong decay widths with/without strangeness suppression mechanism are present in Table 1. In the case of nucleon resonances, we can observe both calculations can be compatible with

Table 3. The pair creation rates and the strangeness suppression factor in the proton.

Ratio	UQM [23]	Exp.	Ref.
$s\bar{s}/d\bar{d}$	0.265	0.22 ± 0.07	[34]
$u\bar{u}/d\bar{d}$	0.568	0.74 ± 0.18	[34]
$2s\bar{s}/(u\bar{u}+d\bar{d})$	0.338	0.25 ± 0.08	[34]
		0.29 ± 0.02	[35]

the experimental data due to the experimental values do not have enough precision. For the case of Δ resonances the strangenes suppression mechanism is beneficial, but for the $\Sigma^*(2030) \to \Xi K$ process, the suppression mechanism is not enough to reproduce the experimental data. Thus in general the suppression mechanism seems to be beneficial, but more precision in the experimental data is needed, and other decay channels should be studied to make a definitive conclusion.

Regarding to electro-production, in the UQM the ratios for exclusive two-body production can be determined in a straightforward way, and Table 2 shows that the observed rates are reproduced very well by our calculation. Here, the isospin symmetry is still valid, thus the calculated ratio $p \to p\pi^0/p \to n\pi^+ = 1/2$ is a consequence of this symmetry.

The calculation of the strangeness suppression factor, $\lambda_s = 2s\bar{s}/(u\bar{u}+d\bar{d})$, takes into account all channels involving pseudoscalar mesons (π, K, η and η') in combination with octet and decuplet baryons. The results are presented in Table 3. The value λ_s is in good agreement with both the values determined in exclusive reactions [34] and in high-energy production [35].

In conclusion, the observed ratios for the production of baryon-meson channels in exclusive reactions can be understood in a simple and transparent way in the framework of the UQM. It is important to emphasize that the UQM results do not depend on the strength of the $^3P_0$ quark-antiquark pair creation vertex. The value of the remaining coefficient (F^2) was taken from previous work, no attempt was made to optimize their values. Finally, the UQM value for the strangeness suppression factor in the proton is in good agreement with the value determined in exclusive reactions [34] as well as the result from high-energy production [35].

We point out the difference between the strangeness suppression mechanism in the strong decays and the strangeness suppression factor extracted in the production rates, in the first case the mechanism is incorporated to take into account the $SU(3)$ symmetry breaking due to the heavier s-mass quark in comparison with the mass of u and d quarks. In the other hand, the strangeness suppression factor obtained from production rates is a consequence of the extra components in the proton wave function treated in similar way as the asymmetry in the proton within the UQM framework[23].

Acknowledgments

This work is supported in part by INFN sezione di Genova.

References

[1] Olive K A *et al.* [Particle Data Group Collaboration], 2014 *Chin. Phys. C* **38** 090001.
[2] Capstick S 1992 *Phys. Rev. D* **46** 2864.
[3] Capstick S and Roberts W 1993 *Phys. Rev. D* **47** 1994; 1994 *Phys. Rev. D* **49** 4570; 1998 *Phys. Rev. D* **58** 074011.
[4] Okubo S 1963 *Phys. Lett.* **5** 165; Iizuka J 1966 *Prog. Theor. Phys. Suppl.* **37** 21.
[5] Bijker R, Ferretti J, Galatà G, García-Tecocoatzi H and Santopinto E 2016 *Phys. Rev. D* **94** 074040.
[6] Micu L 1969 *Nucl. Phys. B* **10** 521.
[7] Le Yaouanc A, Oliver L, Pene O and Raynal J -C 1973 *Phys. Rev. D* **8** 2223; 1974 *Phys. Rev. D* **9** 1415.

IOP Conf. Series: Journal of Physics: Conf. Series **876** (2017) 012010 doi:10.1088/1742-6596/876/1/012010

[8] Eichten E, Gottfried K, Kinoshita T, B Kogut J, Lane D K and -M Yan T 1975 *Phys. Rev. Lett.* **34** 369 ; Eichten E, Gottfried K , Kinoshita T, D Lane K and -M Yan T 1978 *Phys. Rev.* D **17** 3090; 1980 *Phys. Rev.* D **21** 203.
[9] Isgur N and Karl G 1978 *Phys. Rev.* D **18** 4187; 1979 *Phys. Rev.* D **19**, 2653 [1981 Erratum-ibid. **23** 817]; 1979 *Phys. Rev.* D **20** 1191.
[10] Godfrey S and Isgur N 1985 *Phys. Rev.* D **32** 189.
[11] Capstick S and Isgur N 1986 *Phys. Rev.* D **34** 2809 .
[12] Ferraris M, Giannini M M, Pizzo M, Santopinto E and Tiator L 1995 *Phys. Lett.* B **364** 231; Santopinto E, Iachello F and Giannini M M 1998 *Eur. Phys. J.* A **1** 307 ; Santopinto E and Giannini M M 2012 *Phys. Rev.* C **86** 065202; Giannini M M, Santopinto E 2015 *Chin. J. Phys.* **53** 020301; Aiello M 1996 *et al. Phys.Lett.* B 387, 215; Aiello *et al.* 1988*J. of Phys.* G **24** 753; Bijker R, Iachello F, Santopinto E 1988 *J. of Phys.* A **31** 9041; M. De Sanctis *et al.* 2007 *Phys. Rev.* C **6** 062201.
[13] Glozman L Y, Riska D O 1996 *Phys. Rept.* **268** 263; Glozman L Y, Plessas W, Varga K, Wagenbrunn R F 1998 *Phys. Rev.* D **58** 094030.
[14] Loring U, Metsch B C and Petry H R 2001 *Eur. Phys. J.* A **10** 395.
[15] Santopinto E 2005 *Phys. Rev.* C **72** 022201; Ferretti J, Vassallo A and Santopinto E 2011 *Phys. Rev.* C **83**, 065204; De Sanctis M, Ferretti J, Santopinto E and Vassallo A 2016 *Eur. Phys. J.* A **52** 121. De Sanctis, M, Ferretti J 2011 Santopinto E *et al. Phys. Rev.* C **84** 055201, Santopinto E and Ferretti J 2015 *Phys. Rev.* C **92** 025202.
[16] Galata G and Santopinto E 2012 *Phys. Rev.* C **86** 045202.
[17] Bijker R, Iachello F and Leviatan A 1994 *Ann. Phys.* (N.Y.) **236** 69; 1996 *Phys. Rev.* C **54** 1935; 1997 *Phys. Rev.* D **55** 2862; 2000 *Ann. Phys.* (N.Y.) **284** 89.
[18] van Beveren E, Dullemond C and Rupp G 1980 *Phys. Rev.* D **21** 772 [1980 Erratum-ibid. D **22** 787].
[19] van Beveren E, Rijken T A, Metzger K, Dullemond C, Rupp G and Ribeiro J E 1986 *Z. Phys.* C **30** 615.
[20] Ono S and Törnqvist N A 1984 *Z. Phys.* C **23** 59; Heikkila K, Ono S and Törnqvist N A 1984 *Phys. Rev.* D **29** 110 [1984 Erratum-ibid. **29** 2136]; Ono S, Sanda A I and Törnqvist N A 1986 *Phys. Rev.* D **34** 186.
[21] Törnqvist N A and Zenczykowski P1984 *Phys. Rev.* D **29** 2139; 1986 *Z. Phys.* C **30** 83; Zenczykowski P1986 *Annals Phys.* **169** 453.
[22] Tornqvist N A 1995 *Z. Phys. C* **68** 647.
[23] Santopinto E, Bijker R and García-Tecocoatzi H 2016 *Phys. Lett. B* **759** 214 [arXiv:1601.06987].
[24] Ferretti J, Galatà G and Santopinto E 2013 *Phys. Rev. C* **88** 015207.
[25] Ferretti J, Galatà G, Santopinto E,and Vassallo A 2012 *Phys. Rev. C* **86** 015204.
[26] Ferretti J and Santopinto E 2014 *Phys. Rev. D* **90** 094022.
[27] Ferretti J, Galatà G and Santopinto E 2014 *Phys. Rev. D* **90** 054010.
[28] Kalashnikova Y S 2005 *Phys. Rev. D* **72** 034010.
[29] Geiger P and Isgur N 1997 *Phys. Rev. D* **55** 299.
[30] Bijker R, Ferretti J and Santopinto E 2012 *Phys. Rev. C* **85** 035204.
[31] Santopinto E and Bijker R 2010 *Phys. Rev. C* **82** 062202.
[32] Santopinto E and Bijker R 2008 *Few Body Syst.* **44** 95.
[33] Bijker R and Santopinto E 2009 *Phys. Rev. C* **80** 065210.
[34] Mestayer M D *et al.* 2014 *Phys. Rev. Lett.* **113** 152004
[35] Bocquet G *et al.* 1996 *Phys. Lett. B* **366** 447

IOP Conf. Series: Journal of Physics: Conf. Series **876** (2017) 012011 doi:10.1088/1742-6596/876/1/012011

Effect of breakup and resonance states of ^{6}Li on fusion

A. Gómez Camacho

Departamento de Aceleradores. Instituto Nacional de Investigaciones Nucleares, Apartado Postal 18-1027, C.P. 11801, México, D.F.

Abstract. CDCC calculations of total fusion cross sections are presented for reactions of the weakly bound ^{6}Li with targets ^{144}Sm and ^{154}Sm at energies around the Coulomb barrier. Couplings to the low-lying excited states 2^+, 3^- of ^{144}Sm and 2^+, 4^+ of ^{154}Sm are included in the calculations. In the cluster structure frame of $^6\text{Li}\rightarrow \alpha + d$, short range absorption potentials are considered for the interactions between α- and d- with the target to account for fusion. The effect of the excited states of the target on total fusion is investigated, as well as, that from couplings of resonance states of ^{6}Li, namely, $l = 2, J^\pi = 3^+, 2^+, 1^+$. The latter effect is calculated by two different approaches, (a) by considering only resonance states couplings and (b) by omitting these states from the full discretized energy space. Among other things, it is found that resonance and non-resonance continuum couplings give rise to small and similar fusion suppression at the higher energies.

1. Introduction

Lately, reaction mechanisms involving weakly bound nuclei, both stable and radioactive, have been a subject of strong research [1, 2, 3, 4, 5]. Among the most interesting and studied subjects is the effect of breakup of the weakly bound projectile on elastic scattering and fusion reaction processes. The characteristic low binding energy associated to this type of projectiles affects reaction mechanisms mainly in two ways. Namely, the static effect due to the large diffusivity of the projectile matter density. This stretched density lowers the Coulomb barrier and hence enhance fusion. On the other side, the high projectile breakup probability produces strong repulsive couplings that affect elastic and fusion mechanisms. The repulsive polarization potentials so produced, are most important at energies around the barrier and thus, cause fusion suppression.

Other features of reactions with weakly bound nuclei that have been a topic of experimental and theoretical research are the different fusion mechanisms that take place, that is, complete and incomplete fusion. Complete fusion (CF) can be direct complete fusion (DCF) which is a process similar to fusion between strongly bound nuclei, that is, fusion that occurs without a previous excitation of breakup channels. Also, the Sequential complete fusion (SCF) that consists of fusion of all the projectile fragments after a previous breakup. From an experimental point of view, it is very difficult to distinguish the evaporation products from a compound nucleus produced by DCF or SCF, for that reason, only the sum of CF and SCF can be measured. On the other hand, another reaction mechanism is the incomplete fusion (ICF), which is the partial absorption of some fragments while others fly away to the continuum. Total fusion TF, commonly the quantity measured, is the sum of the complete and incomplete fusion. Recently the separate experimental determination of complete and incomplete fusion has been achieved for some reactions with light weakly bound projectiles on medium mass and heavy targets. This has been possible since, contrary to the case of light targets, the compound nucleus decays by the emission of uncharged particles [6, 7, 8]. Another process with weakly bound projectiles is the elastic breakup (EB), in which none of the fragments after breakup is absorbed by the target. It is now well known that in reactions with weakly bound nuclei, couplings from the elastic channel to continuum breakup states, as well as, continuum-continuum couplings have a profound effect on elastic and fusion mechanisms. To study these effects, a complete theoretical description should include all of the aforementioned reaction processes in a single calculation. Presently, the most powerful theoretical tool to do this type of calculations is the Continuum Discretized Coupled-Channel (CDCC) model[9, 10, 11]. This approach has been used to a large range of weakly bound nuclear systems, for instance; the $2n$-halo ^{6}He projectile on ^{59}Co and ^{208}Pb [12, 13], ^{6}Li with targets ^{28}Si, ^{59}Co, ^{58}Ni, ^{144}Sm and ^{208}Pb [9, 12, 14, 15, 16, 17, 18, 19]. Also, ^{7}Li with ^{28}Si [20, 21] and ^{144}Sm [22], the n-halo ^{11}Be with ^{208}Pb [23] and the p-halo ^{8}B with ^{58}Ni [16, 24], ^{12}C [25] and ^{208}Pb [26].

Due to the experimental difficulty to perform separate measurements of complete and incomplete fusion cross sections, most of these studies have considered the effect of breakup couplings on either elastic scattering or total fusion processes. However, recently measurements of CF and ICF have been reported for some weakly bound nuclei with spherical and deformed medium mass targets. For instance, ^{6}Li with targets ^{90}Zr [27], ^{144}Sm and ^{152}Sm [28, 29] and also, the projectile ^{9}Be with ^{181}Ta [30], ^{169}Tm and ^{187}Re [31]. Similarly, theoretical studies that account for the effect of breakup on complete fusion have been performed for a number of systems, see Refs. [32, 33]. In a recent article [34], measurements of complete and incomplete fusion cross sections for the weakly bound ^{6}Li projectile with the highly deformed target ^{154}Sm at energies above the barrier are presented. Also, in this article, coupled-channel calculations are used to study the effect on complete fusion due to couplings to inelastic excited states of ^{154}Sm. As expected, it was found that, this effect is increasingly important as the energy decreases towards the barrier. A comparison of these results with those for the spherical ^{144}Sm with the same projectile ^{6}Li was also discussed. In the present work, CDCC calculations for these systems show that the effect of couplings to excited states of the spherical isotope ^{144}Sm and the largely deformed ^{154}Sm produce an increasing total fusion enhancement as the collision energy decreases below the barrier. As, it will be shown, this effect is more evident for ^{154}Sm than for ^{144}Sm. So, couplings to inelastic excited states of the target produce attractive polarization potentials that lower the fusion barrier. The strengths of these potentials increase for rotational deformed states. It is interesting to point out that total fusion σ_{TF} being composed by CF and ICF, is in principle unaffected by breakup Refs.[1, 35, 36, 37]. However, some enhancement has been found at energies around and below the barrier for the projectiles ^{6}Li, ^{7}Li with targets ^{59}Co and ^{209}Bi [38]. In the calculations, we assume the cluster structure of ^{6}Li$\rightarrow \alpha + d$ ($E_{thres} = -1.47$ MeV). Now, since the masses of the fragments α and d are not so different, their center of mass is not strongly directed towards one of them. Therefore, total fusion can not be calculated by the absorption of their center of mass as in other weakly bound nuclei. For instance, as in the case of the projectile $^{11}Be \rightarrow^{10} Be + n$ for which, the absorption of the center of mass of ^{11}Be ensures the capture of the charged core ^{10}Be[39]. To calculate total fusion, the approach given in Ref. [38] is used. That is, two short-range imaginary potentials, inside the $l = 0$ nominal Coulomb barrier, are used to account for absorption of both or any of the fragments α and d by the target. So, complete and incomplete fusion are implicitly accounted for in this way. These imaginary potentials depend on the relative distance between the fragments and the target. In a different calculation, the effect of continuum resonance state couplings of the weakly bound nucleus ^{6}Li on total fusion is also studied. We follow a similar technique as in Refs. [40, 41], where the effect on elastic scattering from couplings to resonance continuum states of ^{6}Li was determined for reactions with targets ^{28}Si, ^{58}Ni and ^{144}Sm. It was found that resonant states have some important effect on elastic scattering at the lowest collision energies and for the lightest targets. As a matter of fact, resonant state couplings produce a net repulsive polarization potential that reduces incoming flux absorption. Following a similar technique, the effect of continuum resonance states $l = 2$, $j^\pi = 3^+, 2^+, 1^+$ of ^{6}Li on total fusion is studied in this work. That is, two approaches are used: i) Omit resonant states from the full discretized breakup space and ii) Consider only the resonance states in the CDCC calculation of fusion, that is, all non-resonance states are omitted.

In section II, a brief description of the CDCC model is given. Section III is dedicated to give a detailed description of the construction of the CDCC discretized energy space for resonance and non-resonance states. A full description and calculations of total fusion cross sections are also given. The results of the effects on fusion from inelastic excited states of the target, as well as, from resonance and non-resonance continuum breakup couplings of the projectile are given as well. Finally, a summary and conclusions are presented in section IV.

2. Brief CDCC description

A complete description of the CDCC method is given in Refs.[9, 10, 11]. Here, only the basic equations that are required to perform our calculations are presented. We consider the two-body cluster structure of ^{6}Li (α-d) with ground state energy $E_{thres} = -1.47$ MeV. The model space for the ground and continuum states of ^{6}Li is that given in Ref. [38]. The wave function for breakup continuum states of ^{6}Li reads as,

$$\psi^P_{lj}(\mathbf{r}, k) = \{Y_{lm_l}(\hat{\mathbf{r}}) \otimes \chi_{I\mu s\sigma}\}_{lj} \frac{\varphi_{lsj}(r, k)}{r}, \tag{1}$$

where the internal wave function of the α-d system is $\chi_{I\mu s\sigma}$, with $I = 0$ and $s = 1$. $\varphi_{lsj}(r, k)$ describes the α-d relative radial motion with asymptotic wave-number k, orbital angular momentum l and total angular momentum j.

The radial continuum states $\varphi_{lsj}(r, k)$ in Eq. (1) are not square-integrable. However, Ref. [9] provides a prescription for constructing square-integrable wave functions known as bin states. A bin state $u^{(i)}_\beta(r)$ is obtained by a superposition of scattering wave functions within a given interval i, of continuum k values, $k_{i-1} < k < k_i$, *i.e.*,

$$u^{(i)}_{\beta=lsj}(r) = \sqrt{\frac{2}{\pi\eta_l}} \int_{k_{i-1}}^{k_i} w_i(k, l) e^{-i\delta_k(l)} \varphi_{lsj}(r, k) dk, \tag{2}$$

where $\delta_k(l)$ are scattering phase-shifts of φ_β and $w_i(k, l)$ are weight functions defined by,

$$\eta_l = \int_{k_{i-1}}^{k_i} |w_i(k, l)|^2 \, dk. \tag{3}$$

Actually, the weight functions w_i associated to non-resonant bin states are usually set as $w_i(k, l) = 1$, while for resonant states $w_i(k, l) = sin[\delta_k(l)]$.

The total wave function of the three-body system (α-d-*target*) is given by,

$$\Psi(\mathbf{R}, \mathbf{r}, \xi) = \sum_q \sum_\beta F_{\beta q}(\mathbf{R}) \psi^P_\beta(\mathbf{r}) \otimes \Phi^T_q(\xi), \tag{4}$$

where $\Phi^T_q(\xi)$, $q = 0, 1, 2, 3..$ correspond to the ground and inelastic states of the target satisfying $H^T \Phi^T_q(\xi) = \varepsilon^T_q \Phi^T_q(\xi)$. $F_{\beta q}(\mathbf{R})$ represents the projectile-target relative motion in the βq-channel.

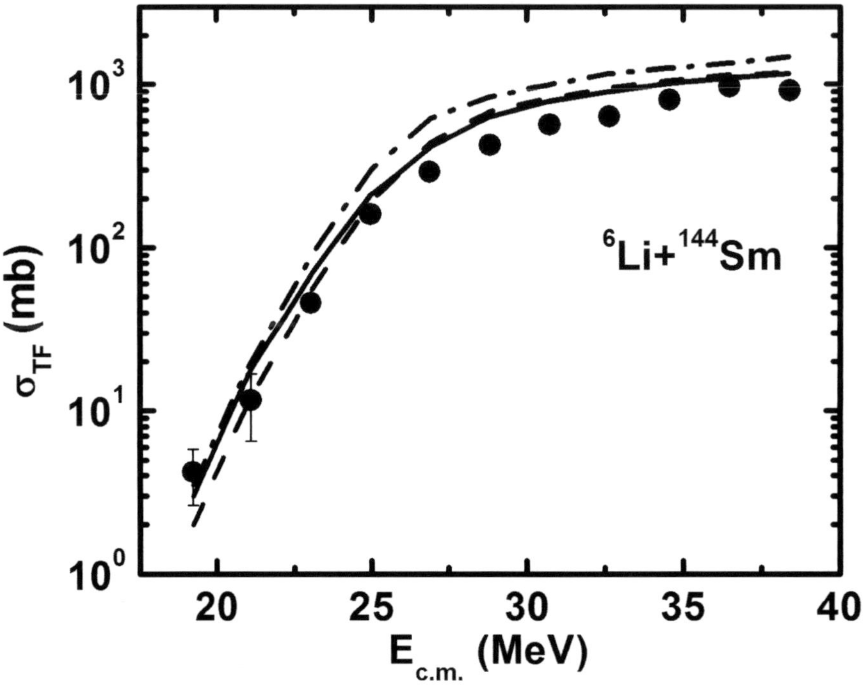

Figure 1. Total fusion cross section for ^{6}Li+^{144}Sm with the full breakup space of ^{6}Li (solid line), without excited states of the target (dashed line) and without breakup states of the projectile (dashed-dotted line).

Projecting the equation of motion of the system onto the projectile states β and onto the excited states of the target q, the following coupled equations are obtained,

$$\left[\hat{T}_R + U^{(J)}_{\beta q,\beta q}(R) - (E - \varepsilon_\beta - \varepsilon^T_q)\right] F^{(J)}_{\beta q}(R) = -\sum_{\beta' q'} U^{(J)}_{\beta q,\beta' q'}(R) F^{(J)}_{\beta' q'}(R). \tag{5}$$

Here, ε_β is the excitation energy of the projectile in the β-state. $U^{(J)}_{\beta q,\beta q}$ and $U^{(J)}_{\beta q,\beta' q'}$ are the radial dependent diagonal and non-diagonal matrices of the interaction potentials $\hat{V}_{dT} + \hat{V}_{\alpha T}$, where $\hat{V}_{dT}$ and $\hat{V}_{\alpha T}$ are the nuclear interactions between the deuteron and α-particle with the target. Notice that $U^{(J)}_{00,00}$ corresponds to the elastic incident channel. As a matter of fact, the interaction potential matrices are given by,

$$U^{(J)}_{\beta q \beta' q'}(R) = < u_\beta \Phi^T_q \left| \hat{V}_{dT}(\mathbf{r}_{dT}, \xi) + \hat{V}_{\alpha T}(\mathbf{r}_{\alpha T}, \xi) \right| u_{\beta'} \Phi^T_{q'} >, \tag{6}$$

where $\mathbf{r}_{dT} = \mathbf{R} - \frac{2}{6}\mathbf{r}$ and $\mathbf{r}_{\alpha T} = \mathbf{R} + \frac{4}{6}\mathbf{r}$ represent the radial distance between the fragments and the target. The integrations in Eq.(6) are carried out over the internal radial coordinate $\mathbf{r}$, the angular coordinates of $\mathbf{R}$ and the internal coordinates of the target ξ.

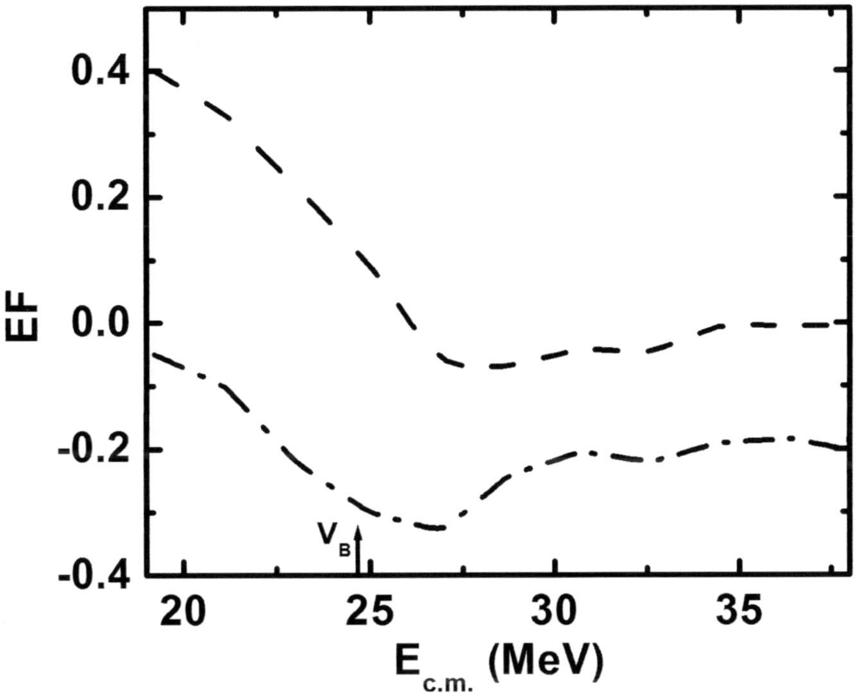

Figure 2. Fusion enhancement and suppression factors from excited states of ^{144}Sm (dashed line) and from breakup states of ^{6}Li (dashed-dotted line).

3. Calculations

3.1. Discretization of continuum space and interaction potentials

The ground state and discretized breakup states of ^{6}Li$\rightarrow d + \alpha$ are constructed by using the interaction given in Ref.[38]. The discretization of the continuum is made as follows. The maximum angular momentum for the relative motion of the fragments is $l_{max} = 3$, larger values do not have any effect on the calculated cross sections. So, bin states are constructed from an initial energy $\varepsilon_0 = 0$ MeV (above the threshold energy $E_{thres} = -1.47$ Mev) up to a maximum energy $\varepsilon_{\rm max} = 6.8$ MeV. For states with $l = 0$, $j^\pi = 1^+$ and $l = 1$, $j^\pi = 0^-, 1^-, 2^-$ an energy step is fixed at $\Delta\varepsilon = 0.5$ MeV. Finer and variable steps are used for resonant states $l = 2$, $j^\pi = 3^+, 2^+$, 1^+, so as to obtain centroid excitation energies and widths close to the corresponding measured values [38]. For bin states $l = 3$, $j^\pi = 4^+, 3^+, 2^+$, a larger step $\Delta\varepsilon = 1.0$ MeV is used. Convergence tests at $\varepsilon_{\rm max} = 7.0, 7.5$ and 8.0 MeV were done with no effect on fusion. Similarly, larger steps $\Delta\varepsilon = 0.75$ and 1.0 MeV were used with no appreciable effect. Coulomb and nuclear potential multipoles are included up to $L_Q = 4$.

As for the nuclear interactions between the fragments α and d and the target of Eq.(6), we use the same potentials as in our previous work for elastic scattering of ^{6}Li+^{144}Sm [41]. That is, the systematic Woods-Saxon potential of Ref. [42] for $\hat{V}_{dT}$ and the density dependent double-folding Sao Paulo potential (SPP) for the $\hat{V}_{\alpha T}$ [43, 44]. These potentials have been modified to account for vibrational and deformed effects of the excited states of the target. In the calculations, the following excited states are included. For the spherical ^{144}Sm, the 2^+ (1160 keV) and 3^- (1810 keV) with $\beta_2 = 0.087$ and $\beta_3 = 0.15$, while for the deformed ^{154}Sm, the 2^+ (82 keV) and 4^+

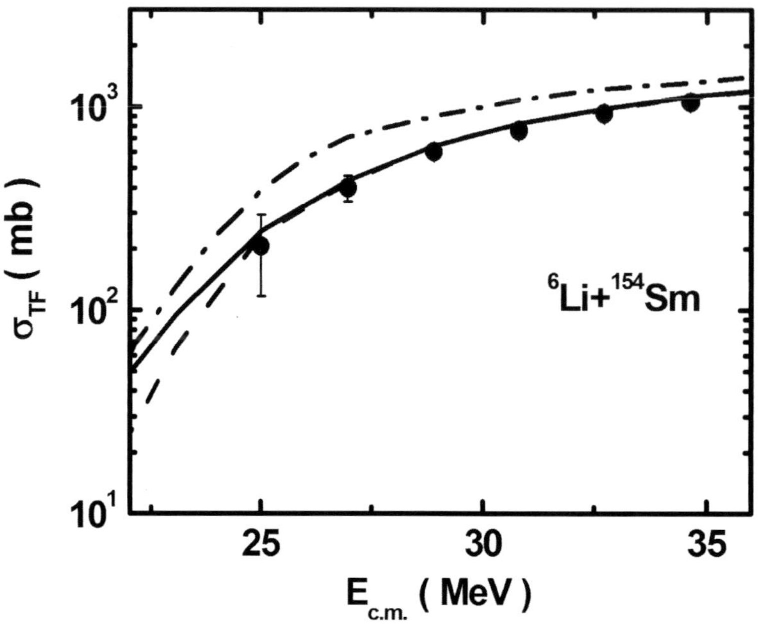

Figure 3. The same as Fig.1 but for ^{154}Sm.

(267 keV) states, with deformation parameters $\beta_2 = 0.34$ and $\beta_4 = 0.08$ [45, 46, 47].

It is important to point out that, the inelastic excitations of the target might be very important, specially in the case of the interaction with the strongly deformed ^{154}Sm target. Because of the limitations of the computational code that will be used to account for the collective degrees of freedom of the targets, we adopted the following approximations. First, we considered that the inelastic and breakup processes of the projectile can not occur at the same time. Then, to calculate the coupling matrix elements $U^J_{\beta=0,\beta'=0}(R)$, the single folding of the sum of the interaction potentials over the projectile g.s. wave function ($< u_0|\hat{V}_{dT} + \hat{V}_{\alpha T}|u_0 >$) is calculated and expanded in multipoles, as in the usual coupled channel calculations.

3.2. Total fusion and effect of resonance and non-resonance states of ^{6}*Li*

CDCC calculations of fusion between the ^{6}Li projectile with targets ^{144}Sm and ^{154}Sm for incident energies around and above the Coulomb barriers, are performed with the code FRESCO [48]. To calculate total fusion cross section σ_{TF}, two short range imaginary Woods-Saxon potentials $W_{\alpha-T}$ and W_{d-T} are considered to account for absorption of the α-particle and *deuteron* by the target. $W_{\alpha-T}$ depends on the radial distance between the α-particle and the target, while W_{d-T} on the radial distance between the *deuteron* and the target. The parameters of these potentials are fixed with strength $W = 25$ MeV, reduced radius $r = 0.9$ fm and diffuseness $a = 0.1$ fm for both ^{144}Sm and ^{154}Sm. Absorption by these potentials happens inside the corresponding Coulomb barriers $V_{B,d} = 13.92$ MeV, $R_{B,d} = 6.4$ fm and $V_{B,\alpha} = 26.5$ MeV, $R_{B,\alpha} = 6.83$ fm respectively. So, the incident flux that passes above or through the barrier, that is inside the short-range potential contributes to fusion. For the system ^{6}Li+^{144}Sm, Fig.1 shows the results for σ_{TF} (solid-line), the calculation without couplings from excited states of the target (dashed-line) and the calculation for the elastic channel, i.e., without couplings to continuum breakup states of the projectile or excited states of the target (dashed-dotted line). These results show

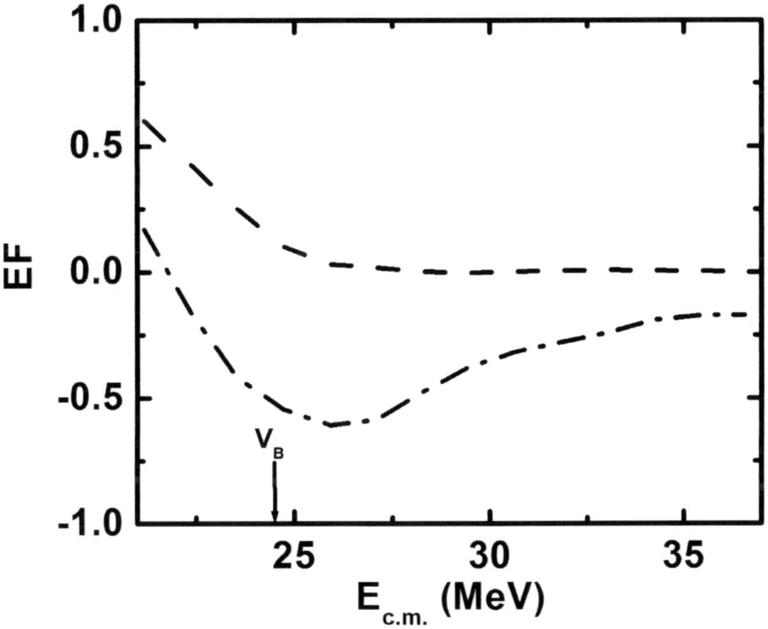

Figure 4. The same as Fig.2 but for ^{154}Sm.

that the calculated total fusion is above the data for energies above the barrier, however there is better agreement around and below the barrier. The total fusion data of Fig.1 are obtained by the sum of complete and incomplete fusion data of Refs. [28, 29]. Fig.2 shows the numerical effect EF on total fusion due to couplings from inelastic excited states of ^{144}Sm (dashed-line) and from continuum breakup states of ^{6}Li (dashed-dotted-line), where the EF is defined by,

$$EF = 1 - \frac{\sigma_{F,n}}{\sigma_{TF}}, \tag{7}$$

here, $n = 1, 2$ for inelastic and breakup couplings respectively. That is, $\sigma_{F,1}$ is the fusion cross section when the inelastic states of the target are omitted, and $\sigma_{F,2}$ when breakup states of the projectile and excited states of the target are not considered. It is observed that couplings to inelastic excited states of ^{144}Sm enhance fusion, particularly for energies below the Coulomb barrier. Here, the enhancement factor EF reaches a value of about 38% at $E_{c.m.} = 20$ MeV. On the other hand, the effect of couplings to breakup states of the projectile produces a net fusion suppression for all energies. The strength of the suppression factor EF increases as the collision energy decreases towards the barrier V_B then to then decrease at lower energies. That is, breakup state couplings produce a net repulsive potential that increases the fusion barrier and therefore reduces fusion. Figs.3 and 4 show the corresponding results for ^{154}Sm. The solid-line of Fig.3 presents the total fusion calculation, while the circles represent the data of Ref. [34]. The dashed-line of Fig. 4 shows that the effect of the excited states of the deformed ^{154}Sm is stronger than for the spherical ^{144}Sm at energies below the barrier. The enhancement factor EF reaches a value of about 50% at $E_{c.m.} = 22$ MeV, while for ^{144}Sm is about 30%. The dashed-dotted line of Fig. 3, shows that breakup state couplings of the projectile are essential to fit the data. This can be seen in Fig. 4, where is observed that these couplings produce a net repulsive effect that rises the barrier. On the other hand, breakup state couplings are more important for ^{154}Sm than for ^{144}Sm as shown by the dotted-dashed lines of Figs. 2 and 4.

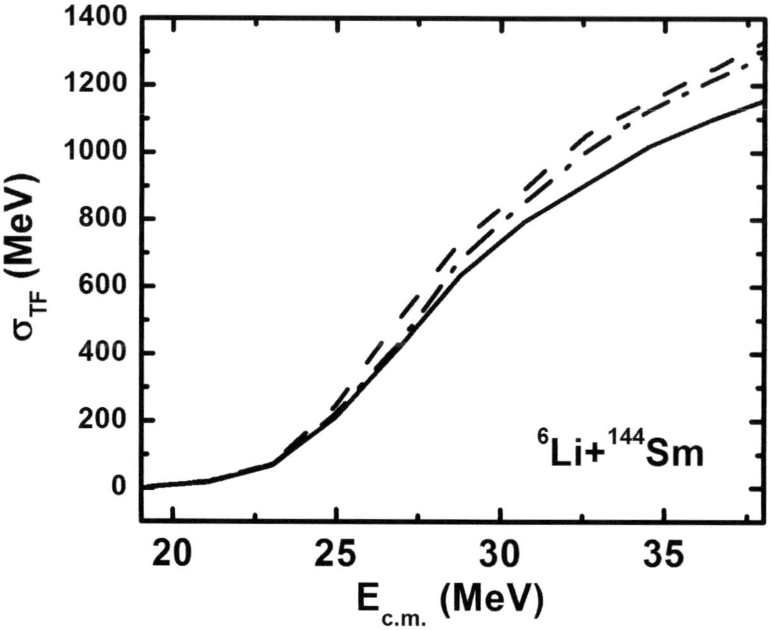

Figure 5. Effect of resonance breakup states of ^{6}Li on total fusion with ^{144}Sm target. The solid-line represents the calculation with the full discretized breakup space. The dashed-line shows the calculation with couplings among resonance states of ^{6}Li, while the dashed-dotted line for couplings among non-resonance states.

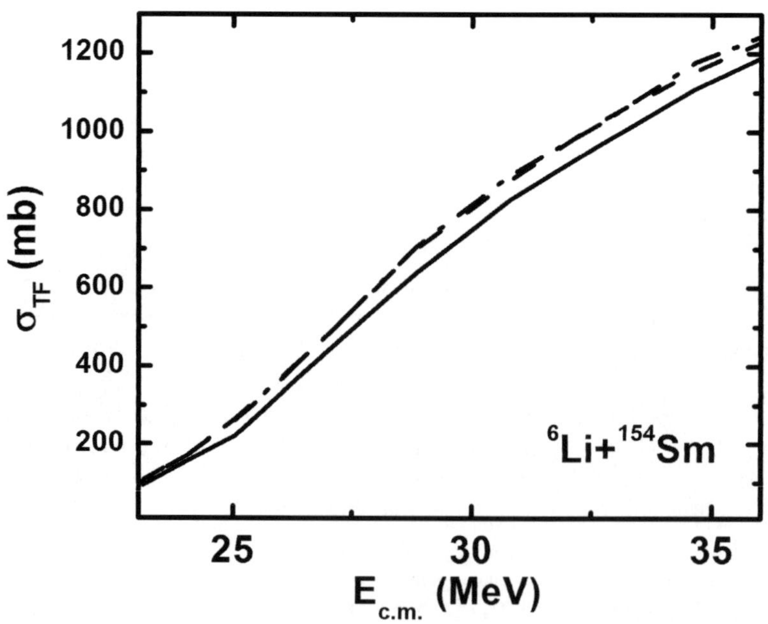

Figure 6. The same as Fig.5 but for ^{154}Sm.

Next, we present the calculations of the effect on total fusion from resonance breakup states $l = 2$, $j^{\pi} = 3^{+}, 2^{+}, 1^{+}$ of the weakly bound nucleus ^{6}Li. The procedure is the same at that used in Refs. [40, 41], where a study of the effect of these resonances on elastic scattering was presented. Two different calculations are performed, namely, fusion is calculated when resonance states are excluded from the full discretized energy space, then when only couplings between the elastic channel and resonance state couplings are considered. The results are shown in Fig.5 for ^{144}Sm and Fig.6 for ^{154}Sm. The solid-lines represent the calculation with the full discretized energy space, the dashed-lines corresponds to the case when only couplings between resonance states are included, and the dashed-dotted lines when resonance states are omitted from the full space. It is observed that, the effect of resonance couplings are more important for ^{144}Sm than for ^{154}Sm. Also, fusion cross section with resonance and non-resonance sub-spaces are higher than the calculation with the full discretized energy space for all the collision energies considered. This implies that the net polarization potentials that appear from separate couplings between resonance and non-resonance states are repulsive. These potentials suppress fusion at all energies. Besides, it is seen that the suppression increases as the collision energy increases, this is more evident for ^{144}Sm than for ^{154}Sm.

4. Summary and conclusions

CDCC calculations of total fusion cross sections have been presented for the nuclear systems ^{6}Li with targets ^{144}Sm and ^{154}Sm at energies around and above the Coulomb barrier. In the CDCC calculations, resonance and non-resonance states of ^{6}Li are discretized up to a maximum energy of 6.8 MeV and discretization steps such that centroid energies of the resonances and widths are close to the experimental values. To account for the effect of excited states of the targets on fusion, low-lying excited states are included in the calculations. The cluster structure of the projectile $^{6}\text{Li} \rightarrow \alpha + d$ is assumed, with global nuclear interactions for the *α-target* and *d-target* sub-systems. For all of the energies studied in this work, it has been shown that calculations with coupling to only the elastic channel are insufficient to fit the fusion data, so, couplings to breakup states of ^{6}Li are very important. On the other hand, the effect of inelastic states of the targets produce attractive polarizations, that in turn, lower the fusion barrier and hence increase fusion. This effect is larger for the deformed ^{154}Sm than for the spherical ^{144}Sm. Short-range imaginary fusion potentials, inside the Coulomb barrier, have been used for the calculation of total fusion. So, complete and incomplete fusion are accounted for, when both fragments are absorbed or only one is captured. On the same footing, the effect on total fusion due to couplings from resonance state couplings of ^{6}Li, namely, $l = 2$, $J^{\pi} = 3^{+}, 2^{+}$ and 1^{+} has been calculated. This effect has been calculated by following two approaches, (a) by omitting the states corresponding to the resonances from the whole energy discrete space and (b) by considering only resonance state couplings. The effects of resonance and non-resonance couplings are stronger for ^{144}Sm than for ^{154}Sm. This is due to the nuclear potentials used for the interactions between the fragments α and d with ^{144}Sm, which are less attractive than those for ^{154}Sm. So, since the Coulomb interaction is the same for both targets, the effect of repulsive couplings from resonance and non-resonance states is more apparent for ^{144}Sm than for ^{154}Sm.

Acknowledgments

Financial support: A. Gómez Camacho from CONACYT, México.

References

[1] L.F. Canto, P.R.S. Gomes, R. Donangelo, M.S. Hussein, Phys. Rep., **424**, 1 (2006).
[2] J.F. Liang, C. Signorini, Int. J. Mod. Phys. E **14**, 1121 (2005).
[3] N. Keeley, R. Raabe, N. Alamanos, J.L. Sida, Prog. Part. Nucl. Sci. **59**, 579 (2007).
[4] L.F. Canto, P.R.S. Gomes, R. Donangelo, J. Lubian M.S. Hussein, Phys. Rep., **596**, 1 (2015).

[5] P.R.S. Gomes *et al.*, *few-body Systems* **57**, 165 (2016).
[6] M. Dasgupta *et al.*, Phys. Rev. Lett. **82**, 1395 (1999), Phys. Rev. C **66**, 04602(R) (2002), Phys. Rev. C **70**, 024606 (2004).
[7] C. Signorini *et al.*, Eur. Phys. J. A **5**, 7 (1999).
[8] P.R.S. Gomes, *et al.*, Phys. Lett. **B634**, 356 (2006), Phys. Rev. C **73**, 064606 (2006).
[9] Y. Sakuragi, M. Yahiro, M. Kamimura, Prog. Theor. Phys. Suppl. **89**, 1 (1986).
[10] Y. Sakuragi, M. Yahiro, M. Kamimura, Prog. Theor. Phys. **70**, 1047 (1983).
[11] N. Austern, Y. Iseri, M. Kamimura, M. Kawai, G. Rawitscher, M. Yahiro, Phys. Rep. **154**, 125 (1987).
[12] C. Beck , N. Keeley , A. Diaz-Torres, Phys. Rev. C **75**, 054605 (2007).
[13] J.P. Fernandez-Garcia, *et al.*, Phys. Lett. B **693**, 310 (2010).
[14] K. Zerva *et al.*, Phys. Rev. C **80**, 017601 (2009).
[15] K. Rusek, N. Keeley, A. Pakou, A. Alamanos, Nucl. Phys. A **784**, 13 (2007).
[16] N. Keeley, R.S. Mackintosh, C. Beck, Nucl. Phys. A **380**, 1 (2010).
[17] N. Keeley, K. Rusek, Phys. Lett. B, **375** (1996).
[18] D.R. Otomar, J. Lubian, P.R.S. Gomes, Eur. Phys. J. A. **46**, 285 (2010).
[19] G.R. Keeley *et al.*, Phys. Rev. C **63**, 024601 (2000).
[20] A. Pakou, *et al.*, Phys. Rev. C **69**, 054602 (2004).
[21] A. Pakou, *et al.*, Eur. Phys. J A **39**, 187 (2009).
[22] D.R. Otomar, *et al.* Phys. Rev. C **80** 034614 (2009).
[23] L.F. Canto, J. Lubian, P.R.S. Gomes, M.S. Hussein, Phys. Rev. C **80** 047601 (2009).
[24] J. Lubian, T. Correa, E.F. Aguilera, L.F. Canto, A. Gomez Camacho, E.M. Quiroz, P.R.S. Gomes, Phys. Rev. C **79**, 064605 (2009).
[25] B. Paes, J. Lubian, P.R.S. Gomes, V. Guimaraes, Nucl. Phys. A **890**, 1 (2012).
[26] J. Rangel, J. Lubian, L.F. Canto, P.R.S. Gomes, Phys. Rev. C **93**, 054610 (2016).
[27] S.P. Hu, *et al.*, Phys. Rev. C **91**, 044619 (2015).
[28] P.K. Rath, S. Santra, N.L. Singh, R. Tripathi, V.V. Parkar, B.K. Nayak, K. Mahata, R. Palit, Suresh Kumar, S. Mukherjee, S. Appannababu, R.K. Choudhury, Phys. Rev. C **79**, 051601 (2009).
[29] P.K. Rath, *et al.*, Nucl. Phys. A **874**, 14 (2012).
[30] N.T. Zhang, *et al.*, Phys. Rev. C **90**, 024621 (2014). 1121 (2005).
[31] Y.D. Fang, *et al.*, Phys. Rev. C. **91**, 014608 (2015).
[32] Bing Wang, Wei-Juan Zhao, P.R.S. Gomes, En-Guang Zhao, Shan-Gui Zhou, Phys. Rev. C **90**, 034612 (2014).
[33] Bing Wang, Wei-Juan Zhao, A. Diaz-Torres, En-Guang Zhao, Shan-Gui Zhou, Phys. Rev. C **93**, 014615 (2016).
[34] M.F. Guo, G.L. Zhang, P.R.S. Gomes, J. Lubian, E. Ferioli, Phys. Rev. C **94**, 044605 (2016).
[35] L.F. Canto, P.R.S. Gomes, J. Lubian, L.C. Chamon, E. Crema, Nucl. Phys. A. **821**, 51 (2009).
[36] L.F. Canto, P.R.S. Gomes, J. Lubian, L.C. Chamon, E, Crema, J. Phys. G. **36**, 015109 (2009).
[37] P.R.S. Gomes, J. Lubian, L.F. Canto, Phys. Rev. C. **79**, 027606 (2009).
[38] A. Diaz-Torres, I.J. Thompson and C. Beck, Phys. Rev. C **68**, 044607 (2013).
[39] A. Diaz-Torres and I.J. Thompson, Phys. Rev. C **65**, 024606 (2002).
[40] A. Gómez Camacho, A. Diaz-Torres, P.R.S. Gomes and J. Lubian, Phys. Rev. C **91**, 014607 (2015).
[41] A. Gómez Camacho, A. Diaz-Torres, P.R.S. Gomes and J. Lubian, Phys. Rev. C **93**, 024604 (2016).
[42] H. An and C. Cai, Phys. Rev. C **73**, 054605 (2006).
[43] L.C. Chamon *et al.*, Phys. Rev. Lett. **79**, 5218 (1997).
[44] L.C. Chamon *et al.*, Phys. Rev. C **66**, 014610 (2002).
[45] S. Raman, C.W. Nestor Jr, P. Tikkanem, At. Data Nucl. Data Tables **78**, 1 (2001).
[46] T. Kibedi, R.H. Spear, At. DataNucl. Data Tables **80**, 35 (2001).
[47] J.R. Leigh *et al.*, Phys. Rev. C **52**, 3151 (1995).
[48] I.J. Thompson, Comput. Phys. Rep. **7,** 167 (1988).

IOP Conf. Series: Journal of Physics: Conf. Series **876** (2017) 012012 doi:10.1088/1742-6596/876/1/012012

In search of a broader microscopic underpinning of the potential energy surface in heavy deformed nuclei

P. O. Hess[1], M. Ermamatov[2]

[1]Instituto de Ciencias Nucleares, UNAM, Circuito Exterior, C.U., A.P. 70-543, 04510 México D.F., Mexico
[2]Instituto de Fisica, Universidade Federal Fluminense, 24210-340, Niteroi, Rio de Janeiro, Brazil
and
Theoretical Physics Department, National University of Uzbekistan, Tashkent, Uzbekistan

E-mail: [1]hess@nculeares.unam.mx , [2]mirshod.ermamatov@if.uff.br

Abstract. Starting from the content of the shell model space and using a simple symplectic as a weight Hamiltonian, the relative positions of different symplectic irreducible representations are deduced. Applying a geometrical mapping leads to a microscopically derived *Potential-Energy-Surface*. After smoothing this surface and fitting a mass parameter to the first excited 6^+-state in the ground state band, the spectrum of a nucleus can be reproduced qualitatively. The method is also used to obtain a first estimation of the quadrupole Potential Energy Surface of any nucleus, allowing to obtain information about the structure of the nucleus in question. Of special interest is the prediction of the structure of nuclei away from the valley of stability and of super-heavy nuclei. The method will be illustrated at ^{184}W. One objective is to show that the *Pauli Exclusion Principle* is the main driving force for the structure of a nucleus, though some further microscopic input has to be used.

1. Introduction

The main objective is to show that by a minimal number of assumptions the *Potential Energy Surface* (PES) of quadrupole deformations of a nucleus can be guessed and that the resulting spectrum is near to the observed one. This enables us also to obtain a first hint to the collective quadrupole structure of super-heavy nuclei and of nuclei not measured yet.

The minimal assumptions are:

1) The use of the shell model space, sub-divided into irreducible representations (irreps) of the symplectic group [1]. The result is the consequence of the *Pauli Exclusion Principle* and up to here the procedure is fully microscopic.

2) The use of a simple test-Hamiltonian, which traces the relative distance in energy of the symplectic irreps. Here enters the importance of the quadrupole.quadrupole interaction, which drives the deformation. This test-Hamiltonian only serves to set the relative weight, position in energy, for each symplectic irrep. The test-Hamiltonian by no means is a microscopic Hamiltonian with all possible contributions needed for a full treatment. As we will see, some

important microscopic information will be hidden in two parameters, namely in the experimental deformation of the nucleus and in the mass parameter.

This alone will enable us to get a first guess on the PES. Further steps are:
3) Because we have not the possibility to diagonalize the Hamiltonian with an arbitrary potential, one has to obtain a PES within the *General Collective Model* (GCM) [2, 3], fitting it to the microscopically derived PES, which will introduce errors.
4) If a state within the ground state band is known, fitting the mass parameter B_2 [2] to this state, if possible with a large angular momentum. The parameter B_2 will adjust the moment of inertia. This parameter hides the contribution from, e.g., the nucleons in the intruder levels.

One main objective of this contribution is that the observation of the *Pauli Exclusion Principle* is the most important ingredient to obtain the information about the structure of a nucleus. Knowing the number of protons/neutrons in their valence shell determines which $SU(3)$ irreps are possible and, thus, the structure of the nucleus.

In this contribution we will illustrate the method at the example of ^{184}W and at ^{252}No.

The paper is organized as follows: In Section 2 we will present the sketch of the procedure, in section 3 it will be applied to ^{184}W and in Section 4 conclusions will be drawn.

2. The procedure

An important ingredient is the symplectic model of the nucleus [1] and its contracted version (see, for example [4]). More details can be retrieved in [5] on which also this contribution is based. In [6] the main steps on how to obtain the microscopically derived PES is explained for the first time and applied to ^{238}U.

The main points are that for heavy nuclei the symplectic irreps for the proton and neutron part are obtained separately. After that, it suffices to use a linear coupling of the proton and neutron irreps, because those are lower in energy as the non-linear coupled ones (which correspond to giant resonances). As a Hamiltonian, which provides a relative weight to each irrep, we use one from the contracted symplectic model [4]:

$$H = \hbar\omega\tilde{N} - \frac{1}{2}\chi\left[Q^c\cdot Q^c - (Q^c\cdot Q^c)_{shell}\right] \quad , \tag{1}$$

where $\tilde{N}$ is the phonon number operator of the protons plus the neutrons, $\hbar\omega$ is $45A^{1/3}$-$25A^{2/3}$ MeV for light and $41A^{1/3}$ MeV for heavy nuclei. The tilde refers to the use of the pseudo-$SU(3)$-model [7, 8], see also further below. The Q^c_m is the physical quadrupole operator and $(Q^c\cdot Q^c)_{shell}$ is the *trace equivalent part* of the quadrupole-quadrupole interaction, which guarantees that even for large deformation in average the shell structure is conserved [9]. The nucleons are filled into the Nilsson orbitals at a deformation given by [10] or by experiment, thus, microscopic information enters which is treated here as a black box. The χ-parameter is determined such that the lowest symplectic irrep has its minimum at the deformation value used as an input.

The symplectic Hamiltonian alone works well when the nucleus is well deformed, i.e., there is one dominant symplectic irrep to which one can restrict. However, for a general nucleus there are strong mixing of the symplectic irreps, which have to be taken into account, when a full microscopic treatment is applied. Here, we want to show by the results that the test-Hamiltonian is sufficient to get an estimate of the relative influence of the symplectic irreps. The mixing will be implicitly taken into account, as will be mentioned further below.

The mapping to a geometrical potential is performed defining first a vector coherent state [11] $\mid \Phi(\alpha_{2\mu})\rangle$, which is given by an operator being an exponential function in the symplectic operators which raise the number of phonons by two, applied to a lowest weight state of an $\widetilde{SU}(3)$ irrep within the valence shell. The $\alpha_{2\mu}$ defines the quadrupole variable. The geometrical potential is the obtained through the expectation value of the Hamiltonian with respect to this vector coherent state. Note, that the shell model space is well defined and is subject to the *Pauli Exclusion Principle.* The content is only determined by the number of protons and neutrons, distributed along the shell model orbitals. The number of active nucleons for light nuclei is represented by all the nucleons, while for heavy nuclei the pseudo-$SU(3)$ is used [7, 8]. In the pseudo-$SU(3)$ shell model the orbitals are divided into *unique* and normal ones, where the *unique* orbitals are all those with the largest spin in each shell and the *normal* ones are all others. The angular momentum $l = j \pm \frac{1}{2}$, within the standard shell model, are redefined to $\tilde{l} = j \mp \frac{1}{2}$ and the content of a former shell with η oscillation quanta is mapped to a pseudo shell $\tilde{\eta} = \eta - 1$. In order to obtain the information of how many protons and neutrons are occupying the *normal orbitals* one needs the information on the deformation (which contains microscopic information) and fill in the nucleons at this deformation within the Nilsson diagrams from below. This division into two different kinds of orbitals is by no means sufficient. The intruder orbitals are important, for example, to get the moment of inertia right. However, this we treat as a parameter and the effects of the intruder orbitals are implicitly taken into account.

Restricting to *only one* symplectic irrep, in lowest order the potential is given by

$$
\begin{aligned}
V &\approx \frac{65}{16\pi}(\chi N_s^2)\tilde{\beta}_s^2 + \frac{5}{4\pi}N_s\left(\hbar\omega - \frac{7}{6}\chi N_s\right)(\tilde{\beta} - \tilde{\beta}_s)^2 \\
&- (\chi N_s^2)\frac{5}{\pi}\tilde{\beta}_s\tilde{\beta} + \frac{5}{2\pi}N_s\left(\hbar\omega + \frac{5}{6}\chi N_s\right)\tilde{\beta}_s\tilde{\beta}\left(1 - \cos(\gamma - \gamma_s)\right) \quad ,
\end{aligned} \tag{2}
$$

where $\tilde{\beta}_s$ is related to a particular pseudo-$SU(3)$ irrep $(\widetilde{\lambda}, \widetilde{\mu})$ via

$$
\tilde{\beta}_s = \left[\left(\frac{4\pi}{5N_s^2}\right)\left(\widetilde{\lambda}^2 + \widetilde{\lambda}\widetilde{\mu} + \widetilde{\mu}^2 + 3\widetilde{\lambda} + 3\widetilde{\mu}\right)\right]^{\frac{1}{2}} \quad . \tag{3}
$$

The N_s is the total number of phonons in the lowest pseudo-$SU(3)$-state plus $3\frac{(\tilde{A}-1)}{2}$ and $\tilde{\beta} = \frac{\beta}{\kappa}$ ($\tilde{\beta}_s = \frac{\beta_s}{\kappa}$). (In [6] the factor $\frac{1}{\kappa}$ was omitted for β_s for heavy nuclei, which does not lead to significant deviations for prolate nuclei but being important for the case of oblate nuclei.) The value κ is approximately given by $\frac{1}{2}(\frac{\tilde{\eta}_p+1.5}{\tilde{\eta}_p+0.5} + \frac{\tilde{\eta}_n+1.5}{\tilde{\eta}_n+0.5})$ with $\tilde{\eta}_p$ and $\tilde{\eta}_n$ being the pseudo valence shell number for protons and neutrons respectively. $(\tilde{\beta}_s, \gamma_s)$ in (2) is related to the deformation of the symplectic band head in the $0\hbar\omega$ [4] (see also Eq. (3)). The total PES is then obtained by the envelope of all potentials related to the sum of all symplectic irreps. Note, that the determination of the microscopic potential has no free parameters. By taking the envelop, we implicitly assume that the system can change from one potential of the type (2) to another one, when the last becomes lower. In the symplectic model as such, there are no transitions between different potentials. This means that the procedure proposed implies a more general Hamiltonian.

For the GCM collective potential we use [2]

$$
\begin{aligned}
V &= \frac{C_2}{\sqrt{5}}\beta^2 - \sqrt{\frac{2}{35}}C_3\beta^3\cos(3\gamma) + \frac{C_4}{5}\beta^4 - \sqrt{\frac{2}{175}}C_5\beta^5\cos(3\gamma) \\
&+ \left(\frac{2}{35}\right)C_6\beta^6\cos^2(3\gamma) + \frac{D_6}{5\sqrt{5}}\beta^6 \quad .
\end{aligned} \tag{4}
$$

The potential parameters are approximately deduced by fitting (4) to (2).

β_0 [12]	κ	$\widetilde{N}_{\rm vp}$	$\widetilde{N}_{\rm vn}$	$\widetilde{\eta}_{\rm p}$	$\widetilde{\eta}_{\rm n}$	$\widetilde{A}_{\rm p}$	$\widetilde{A}_{\rm n}$	$\widetilde{A}$	$N_{\rm s}$
-0.2344(17)	1.25	14	20	3	4	34	60	94	381.5

Table 1. The values of quantities used in the calculation of microscopic PES ^{184}W. The κ is a scale factor appearing in the mapping of $SU(3)$ to pseudo-$SU(3)$. $\widetilde{N}_{\rm vp}$ are the number of valence protons in the pseudo-valence shell $\widetilde{\eta}_{\rm p}$ and similar for the neurons. $\widetilde{A}_{\rm p}$is the number of protons in the normal orbitals and $\widetilde{A}_{\rm n}$ the same for the neutrons. $\widetilde{A}$ is the total number of nucleons in normal orbitals. Finally, $N_{\rm s}$ is the number of quanta of the normal orbitals in the pseudo-$SU(3)$ plus $\frac{3}{2}(\widetilde{A}-1)$.

C_2	C_3	C_4	C_5	C_6	D_6	B_2
-521.77	-337.80	14306.01	-502.64	1902.26	-60439.94	112.697

Table 2. The values of the phenomenological potential parameters for ^{184}W, determined by fitting.

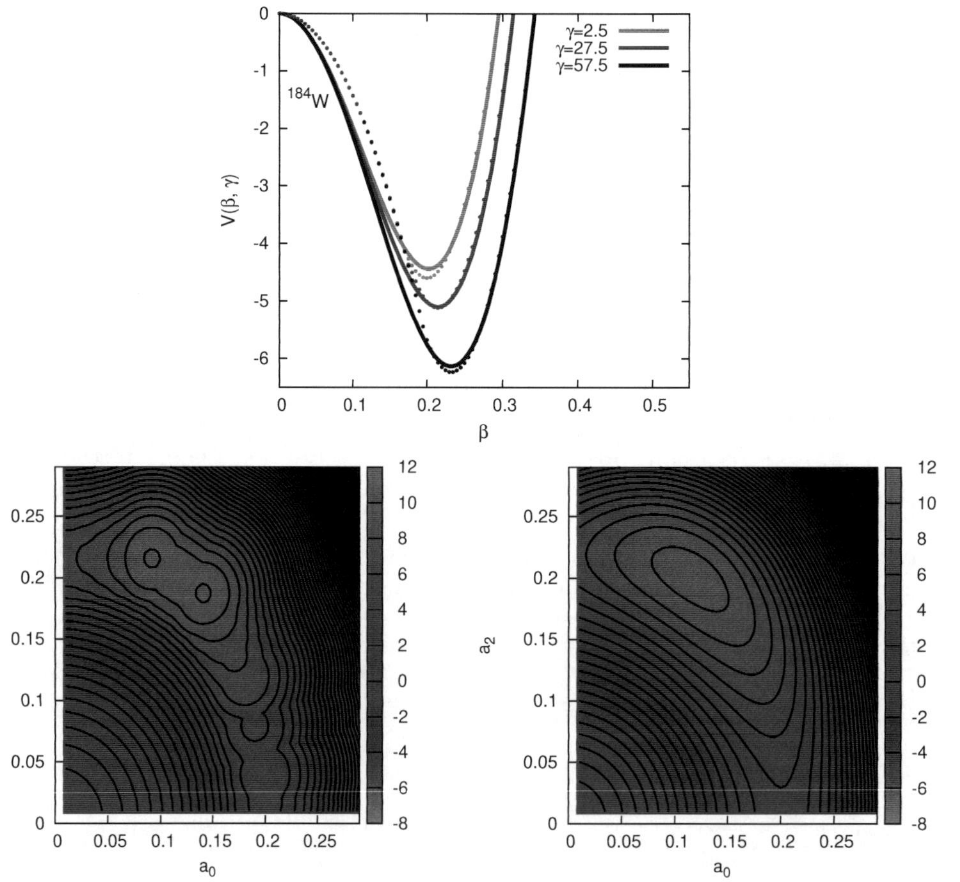

Figure 1. Comparison of microscopic and phenomenological potential energy surfaces of ^{184}W.

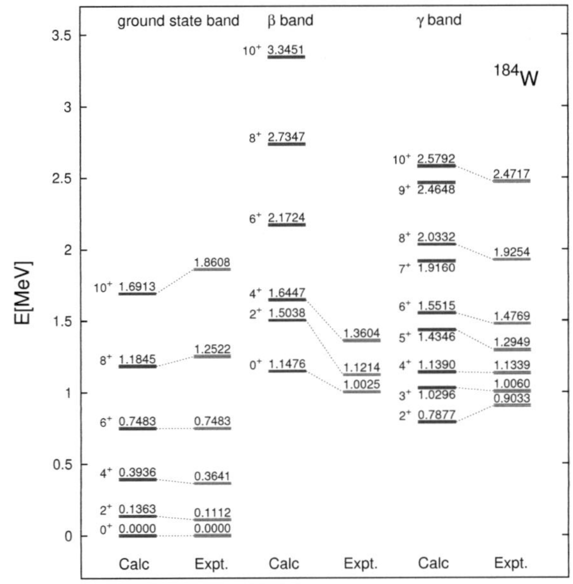

Figure 2. Comparison of the calculated energy levels ^{184}W with experimental data.

$L_{gs} \rightarrow L'_{gs}$	B(E2)		$L_\beta \rightarrow L'_\beta$	B(E2)		$L_\gamma \rightarrow L'_\gamma$	B(E2)	
	Calc	Exp		Calc	Exp		Calc	Exp
$0^+ \rightarrow 2^+$	1.95	3.78(6)	$0^+ \rightarrow 2^+$	1.68	-	$2^+ \rightarrow 3^+$	0.75	-
$2^+ \rightarrow 4^+$	1.06	1.86(5)	$2^+ \rightarrow 4^+$	0.07	-	$2^+ \rightarrow 4^+$	0.59	-
$4^+ \rightarrow 6^+$	1.01	1.63(5)	$4^+ \rightarrow 6^+$	0.55	-	$3^+ \rightarrow 4^+$	0.38	-
$6^+ \rightarrow 8^+$	1.03	1.5(3)	$6^+ \rightarrow 8^+$	0.78	-	$3^+ \rightarrow 5^+$	0.66	-
						$4^+ \rightarrow 5^+$	0.29	-
						$4^+ \rightarrow 6^+$	0.83	-
						$5^+ \rightarrow 6^+$	0.18	-
						$5^+ \rightarrow 7^+$	0.85	-
						$6^+ \rightarrow 7^+$	0.14	-
						$6^+ \rightarrow 8^+$	0.94	-
						$7^+ \rightarrow 8^+$	0.10	-

Table 3. Calculated intraband B(E2) values in e^2b^2 unit with experimental data for ^{184}W. Units are in e^2b^2.

$L_{gs} \to L_\beta$	B(E2)		$L_{gs} \to L_\gamma$	B(E2)		$L_\gamma \to L_\beta$	B(E2)	
	Calc	Exp		Calc	Exp		Calc	Exp
$0^+ \to 2^+$	29.4	-	$0^+ \to 2^+$	222.3	-	$2^+ \to 2^+$	0.9	-
$2^+ \to 2^+$	0.7	-	$2^+ \to 2^+$	175.0	-	$2^+ \to 4^+$	134.8	-
$2^+ \to 4^+$	1.6	-	$2^+ \to 3^+$	114.0	-	$4^+ \to 4^+$	245.7	-
$4^+ \to 4^+$	2.9	-	$2^+ \to 4^+$	17.0	-	$4^+ \to 6^+$	23.2	-
$4^+ \to 6^+$	1.1	-	$4^+ \to 4^+$	165.5	-	$6^+ \to 6^+$	262.4	-
$6^+ \to 6^+$	2.7	-	$4^+ \to 5^+$	60.0	-	$6^+ \to 8^+$	8.9	-
$6^+ \to 8^+$	0.8	-	$4^+ \to 6^+$	5.9	-	$8^+ \to 8^+$	256.3	-
$8^+ \to 8^+$	2.6	-	$6^+ \to 6^+$	154.5	-			-
			$6^+ \to 7^+$	40.2	-			-
			$6^+ \to 8^+$	3.1	-			-
			$8^+ \to 8^+$	145.5	-			-

Table 4. Calculated interband 10^3·B(E2) values in e^2b^2 units for ^{184}W. Units are in e^2b^2.

The procedure dos not provide us within information on ground state energies and related observables of the ground state of a nucleus. However, it can provide information on stability: A spherical PES hints to a particular stable configuration.

3. Applications

Due to lack of space, we simply resume the main results, as the energy spectrum, the structure of the PES and some transition values. The parameter values can be retrieved from the Tables 1 and 2 .

In the presentation of the conference the results for ^{182}W, ^{184}W and ^{186}W were presented. In this written version, only the ^{184}W-isotopes is considered. Experimental data were taken from [12]. This contribution serves to show the basic procedure. Applications to other known nuclei and super-heavy nuclei are published in [5]. The deduced microscopic potential has an immediate consequence on stability considerations of super-heavy nuclei: When the PES is spherical it clearly indicates stability. As can be seen from Figure 1, the GCM potential reproduces satisfactorily the microscopic derived one. In Figure 2 the spectrum is comparable to the experimental one, though the staggering within the sidebands is exaggerated. This was also observed in [5], where many different isotope chains were considered, including in the range of super-heavy nuclei. The agreement to the $B(E2)$ transitions within bands (see Table 2) are quite well predicted. A general feature is the exaggerated staggering within the vibrational bands.

Also a better agreement can be obtained using the deduced microscopic potential and not the approximation to the GCM-potential. However, in this case one needs a numerical routine [13], which is not available anymore.

An improvement in reducing the staggering within the vibrational bands can be achieved introducing a $1/\beta^2$ interaction as done in the Davidson potential [14], which implies however an additional parameter. Currently, we are including this interaction.

Finally, in Fig. 3 we present an example of a super-heavy nucleus. Only the PES is shown, while the spectrum can be retrieved in [5]. Many more have been considered in [5]. The particular example chosen is due to the knowledge of the states within the ground state band, which helps us to determine the mass parameter.

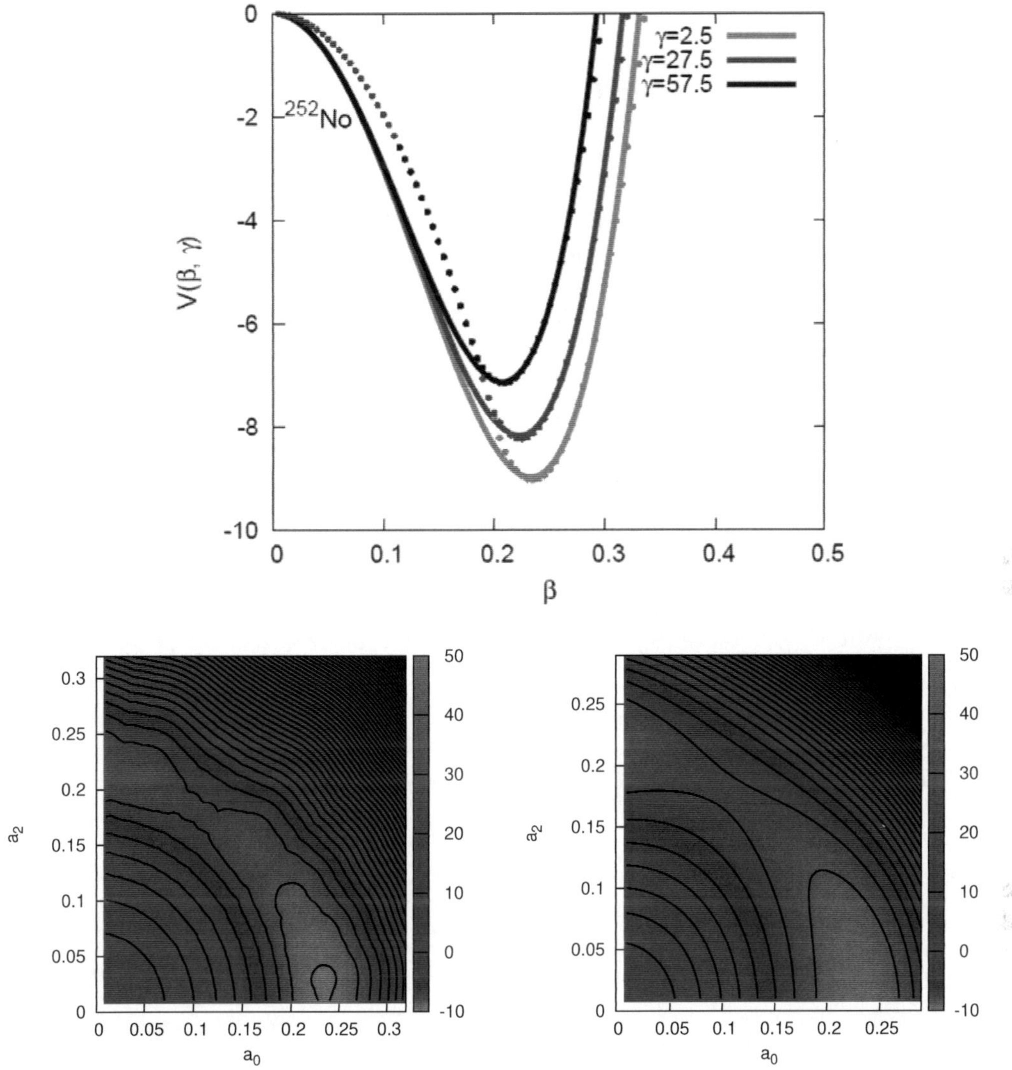

Figure 3. Comparison of microscopic and phenomenological potential energy surfaces of the super-heavy nucleus ^{252}No.

4. Conclusions

We estimated the structure microscopic PES with a minimum of physical assumptions and no free parameters, leading to the prediction of the collective quadrupole structure of nuclei in a wide range of heavy nuclei. The driving force for the structure is the *Pauli Exclusion Principle*, though further microscopic input are hidden in the information of the deformation (taken from experiment or the tables [10]) and the use of a weight Hamiltonian whose origin is the symplectic model. The symplectic model Hamiltonian is only used as a weight to obtain the order of the relative distance of the symplectic irreps, a difference due to the quadrupole-quadrupole interaction.

Finally, for the spectrum one fits the mass parameter to the energy of the first 6^+ state, or one assumes a reasonable value. This also contains information on the microscopic structure, because it fits the moment of inertia.

As an illustrative examples we discussed the ^{184}W-isotope. The spectrum and $B(E2)$ transitions were calculated with a quite good agreement to experiment. A more complete presentation of the procedure was presented in the talk and especially can be found in [5].

Acknowledgment
The authors acknowledge financial help from DGAPA-PAPIIT (IN100315) and CONACyT (Project no. 251817). M.E. acknowledges financial support from the *Brazilian National Counsel of Technology and Scientific Development* (Project No. 165371/2015-3).

References

[1] D.J. Rowe, Rep. Prog. Phys. **48** (1985), 1419;
D.J. Rowe, Prog. Part. Nucl. Phys. **17** (1996), 265.
[2] P.O.Hess, M.Seiwert, J.A.Maruhn, W.Greiner, Zeitschr. f. Phys. **A296** (1980), 147.
[3] P.O.Hess, J.A.Maruhn, W.Greiner, Phys. **G7** (1981), 737.
[4] O. Castaños and J. P. Draayer, Nucl. Phys. A**491** (1989), 349
[5] M. Ermamatov and P. O. Hess, Ann. Phys. (N.Y.) **371** (2016), 125.
[6] O.Castaños, P.O.Hess, P.Rocheford, J.P.Draayer, Nucl. Phys. **A524** (1991),469.
[7] A. Arima, M. Harvey, and K. Shimizu, Phys. Lett. B **30**, (1969), 517.
[8] K. T. Hecht and A. Adler, Nucl. Phys. A **137** (1969), 129.
[9] G. Rosensteel and J. P. Draayer, Nucl. Phys. A**436** (1985), 445.
[10] P. Möller, J. R. Nix, W. D. Myers and W. J. Swiatetecki, Atom. Data and Nucl. Data Tab **59** (1995), 185.
[11] K. T. Hecht, *The vector Coherent Sate Method and its Application to Problems of Higher Symmetries*, Lecture Notes in Physics 290, (Springer, Heidelberg, 1997).
[12] www.nndc.bnl.gov/ensdf.
[13] D.Troltenier, J.A.Maruhn, W.Greiner, P.O.Hess, Zeit. f. Phys. A**343** (1992), 25.
[14] D. Bonatsos, E. A. McCutchan, N. Minkov, R. F. Casten, P. Yotov, D. Lenis, D. Petrellis, and I. Yigitoglu, Phys. Rev. C **76** (2007), 064312.

IOP Conf. Series: Journal of Physics: Conf. Series **876** (2017) 012013 doi:10.1088/1742-6596/876/1/012013

On the investigation of resonances above and below the threshold in nuclear reactions of astrophysical interest using the Trojan Horse Method.

M. La Cognata[a], **C. Spitaleri**[a,b], **S. Cherubini**[a,b], **M. Gulino**[a,c], **I. Indelicato**[a,b], **L. Lamia**[b], **R.G. Pizzone**[a], **S. Romano**[a,b], **O. Trippella**[d,e], **A. Tumino**[a,c]

[a]Laboratori Nazionali del Sud - INFN, Catania, Italy
[b]Dipartimento di Fisica e Astronomia, University of Catania, Catania, Italy
[c]Facoltá di Ingegneria e Architettura, Kore University of Enna, Enna, Italy
[d]Dipartimento di Fisica e Geologia, University of Perugia, Perugia, Italy
[e]Sezione INFN di Perugia, Perugia, Italy

E-mail: `lacognata@lns.infn.it`

Abstract. The occurrence of resonances in reactions of astrophysical interest might significantly enhance the astrophysical factor with respect to the direct reaction contribution, divert nucleosynthesis path and change the energy production, with significant impact on astrophysics. Moreover, the determination of resonance parameters, that is, energy, spin-parity and partial widths, allows one to perform nuclear structure studies leading, for instance, to determine the cluster structure of the state under investigation. However, nuclear reactions in stars take place at energies well below ~ 1 MeV owing to the typical temperatures characterising these environments. Therefore, the Coulomb barrier exponentially suppressing the cross section and the electron screening effect, due to the shielding of nuclear charges by atomic electrons, make it very difficult to provide accurate astrophysical factors. The THM is an indirect method allowing to overcome such difficulties. It makes use of quasi-free reactions with three particles in the exit channel, $a + A \to c + C + s$, to deduce the cross section of the reaction of astrophysical interest, $a + x \to c + C$, under the hypothesis that A shows a strong $x + s$ cluster structure, right at astrophysical energies. By using a generalised R-matrix approach, the resonance parameters can be deduced from THM data allowing one to perform a full spectroscopic study of low-energy and sub-threshold resonances. In this work, we will discuss two examples of reactions of astrophysical interest, whose cross sections show a resonant behaviour: the ^{19}F$(p, \alpha)^{16}$O cross section that displays resonances at energies above the particle emission threshold and the ^{13}C$(\alpha, n)^{16}$O reaction, dominated by the -3 keV sub-threshold resonance due to the 6.356 MeV level in ^{17}O.

1. Introduction

Nuclear astrophysics deals with the investigation of nuclear physics phenomena influencing astrophysical sites such as stars or the early universe. Nuclear reactions power many astrophysical processes like energy production and synthesis of the chemical elements, therefore, cross sections $\sigma(E)$ are among the main input parameters of the codes used to model these scenarios. However, in the case of many astrophysical phenomena, such as quiescent stellar burning, energies of interest are so low that for charged particles the Coulomb barrier strongly diminishes cross sections making the measurement of such cross sections at energies of interest for astrophysics not always possible.

These energies (the so-called Gamow window [1, 2]) usually vary between few keV and few hundreds of keV, therefore cross sections can be much smaller than 1 nb, making extrapolation from high energies the only way to estimate them at the energies of interest. Extrapolation is often performed by means of the astrophysical factor [1, 2]:

$$S(E) = \sigma(E)\, E\, \exp(2\pi\eta)\,, \tag{1}$$

where $\eta = Z_1 Z_2 e^2/\hbar v$ is the Sommerfeld parameter, Z_1 and Z_2 the atomic number of interacting nuclei and v their relative velocity. $S(E)$ is a smoothly varying function of energy because the gross of the Coulomb barrier penetration factor is compensated for owing to the presence of the inverse of the Gamow factor. In the case of resonant reactions, proceeding through an excited state of the intermediate compound nucleus, significant deviations from the smooth behaviour might be expected.

Therefore, extrapolation might introduce systematic errors since the occurrence of unknown or unpredicted resonances might dramatically enhance the astrophysical S(E)-factor, strongly influencing astrophysical models. These considerations apply to both low-energy resonances and sub-threshold resonances as they may produce sizeable modifications of the S-factor due to, for instance, interference with another resonance, especially in the case of broad resonances [1, 2].

Great improvements in the experimental approach used to measure vanishingly small cross sections have been devised, making it possible to extend the measurements of some reactions down to astrophysical energies (see, for instance, [3]). However, approaching interaction energies comparable with the electron binding energies in atoms the presence of atomic electrons cannot be neglected. Electron clouds determine an enhancement of the astrophysical factor related to the shielding of the nuclear charges by the surrounding negatively-charged electrons and not to nuclear interaction (see [1, 2] for a general discussion and [4, 5] for two examples). Even in those few cases when the reaction cross sections were measured inside the Gamow window, electron screening prevented the access to the bare-nucleus cross section, making extrapolation unavoidable.

Indeed, electron screening behaves differently in the laboratory and in astrophysical environments, as in the former projectile and target are in the form of ions and atoms or molecules, while in stars, for instance, matter is in the form of plasma. However, our present understanding of the electron screening effect is rather imperfect as experimental values often exceed theoretical upper limits [6], thus potential systematic errors might be introduced in the evaluation of the bare-nucleus $S(E)$ and, as a consequence, in its extrapolation to low energies.

Because of the problems affecting direct measurements at astrophysical energies, indirect techniques have been developed to bypass them and attain the astrophysical factor at low energies. For instance, the Trojan Horse Method (THM) [7] is a valid technique to get information on the S(E) at astrophysical energies for reactions having charged particles and neutrons in the exit channel, with no Coulomb and centrifugal barrier suppression neither electron screening. In the case of radiative capture reactions, the Asymptotic Normalization Coefficient (ANC) [8] approach has allowed to obtain the direct-capture zero-energy S(E) factor with very high accuracy. The Coulomb dissociation (CD) was also developed [9] to determine the low-energy cross section of charged-particle induced reactions having a γ-ray in the exit channel.

2. The THM for resonant reactions

The THM has been established as a tool to investigate low-energy nuclear reactions by C. Spitaleri [10, 11] following the pioneering work of G. Baur [12]. In more recent years, A.M. Mukhamedzhanov has strongly improved the theoretical formalism [13, 14], introducing

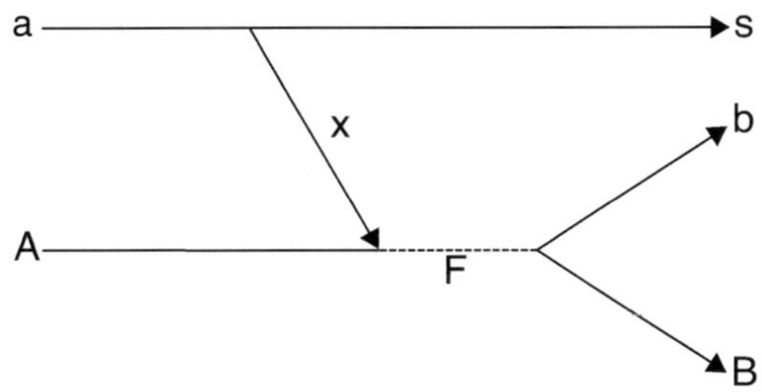

Figure 1. Sketch of the participant (x)-spectator (s) mechanism leading to the population of excited states of the intermediate nucleus F, necessary condition for the application of the THM formalism.

the modified R-matrix approach, allowing for the analysis of multi-resonance reactions with unprecedented accuracy (see, for instance, [15, 16, 17, 18]). These upgrade in the theory has impacted astrophysics, as it is shown in the case of evolved star nucleosynthesis and presolar grains [19, 20]. In the THM general framework, the cross section of the $A(x,b)B$ reaction is obtained through the $A(a,bB)s$ reaction performed at high energies (several tens of MeV), so it is not hindered by the Coulomb and centrifugal barriers of the target-projectile system and no electron screening enhancement alters the trend of the cross section. However, particle x is virtual so the $A(x,b)B$ THM is half-off-energy-shell (HOES) and cannot be simply compared with the corresponding direct (on-energy-shell, OES) cross section [13]. The modified R-matrix approach has been introduced to extract the astrophysical S-factor of interest from the quasi-free (QF) reaction yield, accounting for HOES effects.

Adopting the plane wave approximation (PWA) in the prior form and neglecting, for sake of simplicity, spins of the involved particles, the amplitude of the process $a+A \to b+B+s$ (Fig.1) takes the form:

$$M^{PWA(prior)}(P,\mathbf{k}_{aA}) = \left\langle \chi_{sF}^{(0)}\Psi_{bB}^{(-)}|V_{xA}|\varphi_a\varphi_A\chi_{aA}^{(0)}\right\rangle , \quad (2)$$

where $P=(\mathbf{k}_{sF},\,\mathbf{k}_{bB})$ is the six-dimensional momentum describing the three-body system s, b and B. $\chi_{aA}^{(0)}=\exp\left(i\mathbf{k}_{aA}\cdot\mathbf{r}_{aA}\right)$, $\chi_{sF}^{(0)}=\exp\left(i\mathbf{k}_{sF}\cdot\mathbf{r}_{sF}\right)$, $\mathbf{r}_{ij}$ and $\mathbf{k}_{ij}$ are the relative coordinate and relative momentum of i and j nuclei, $\Psi_{bB}^{(-)}$ is the wave function of the fragments b and B in the exit channel, $F=b+B$, V_{xA} is the interaction potential of x and the target nucleus A, φ_a and φ_A are the bound state wave function of nuclei a and A, respectively. If we assume that resonant reaction mechanism is dominant in the explored energy region, taking into account spins of the interacting particles, considering only the s-wave bound state $a=s+x$ and neglecting the internal degrees of freedom of the transferred particle x, we get the prior PWA amplitude of the THM cross section in the form [14, 7]:

$$\begin{aligned}
M^{PWA(prior)}(P,\mathbf{k}_{aA}) &= (2\pi)^2\sqrt{\frac{1}{\mu_{bB}k_{bB}}}\varphi_a(\mathbf{p}_{sx}) \\
\times \sum_{J_FM_Fj'll'm_{j'}m_lm_{l'}M_x} & i^{l+l'}\left\langle jm_jlm_l|J_FM_F\right\rangle\left\langle j'm_{j'}l'm_{l'}|J_FM_F\right\rangle \\
&\times\left\langle J_xM_xJ_AM_A|j'm_{j'}\right\rangle\left\langle J_sM_sJ_xM_x|J_aM_a\right\rangle \\
&\times\exp\left[-i\delta_{bB\,l}^{hs}\right]Y_{lm_l}(-\hat{\mathbf{k}}_{bB}) \\
&\times\sum_{\nu\tau=1}^{N}\left[\Gamma_{\nu bBjlJ_F}\right]^{1/2}\left[\mathbf{A}^{-1}\right]_{\nu\tau}Y_{l'm'}^{*}(\hat{\mathbf{p}}_{xA}) \\
&\times\sqrt{\frac{R_{xA}}{\mu_{xA}}}\left[\Gamma_{\nu xAl'j'J_F}(E_{xA})\right]^{1/2}P_{l'}^{-1/2}(k_{xA},R_{xA}) \\
\times\,(j_{l'}(p_{xA}R_{xA})&\left[(B_{xA\,l'}(k_{xA},R_{xA})-1)-D_{xA\,l'}(p_{xA},R_{xA})\right] \\
&+2Z_xZ_Ae^2\mu_{xA}\int_{R_{xA}}^{\infty}\mathrm{d}r_{xA}\frac{O_{l'}(k_{xA},r_{xA})}{O_{l'}(k_{xA},R_{xA})}j_{l'}(p_{xA}r_{xA})\Bigg) . \quad (3)
\end{aligned}$$

Here, $\mathbf{p}_{ij}$ is the $i-j$ relative momentum in the case of off-energy-shell particles, thus $E_{ij} \neq p_{ij}^2/2\mu_{ij}$ (while $\mathbf{k}_{ij}$ is calculated assuming the particles on-shell), $\delta_{bB\,l}^{hs}$ is the solid sphere scattering phase shift, R_{xA} the $x+A$ channel radius,

$$B_{xA\,l'}(k_{xA}, R_{xA}) = R_{xA} \frac{\frac{\partial O_{l'}(k_{xA},R_{xA})}{\partial r_{xA}}|_{r_{xA}=R_{xA}}}{O_{l'}(k_{xA}, R_{xA})} \tag{4}$$

is the logarithmic derivative as in the R-matrix method,

$$O_{l'}(k_{xA}, R_{xA}) = \sqrt{\frac{k_{xA}R_{xA}}{P_{l'}(k_{xA}, R_{xA})}} \exp\left[-i\delta_{xA\,l'}^{hs}\right] \tag{5}$$

is the outgoing spherical wave, $P_{l'}(k_{xA}, R_{xA})$ the $l'-$wave penetrability factor,

$$D_{xA\,l'}(p_{xA}, R_{xA}) = R_{xA} \frac{\frac{\partial j_{l'}(p_{xA},R_{xA})}{\partial r_{xA}}|_{r_{xA}=R_{xA}}}{j_{l'}(p_{xA}, R_{xA})} \tag{6}$$

the logarithmic derivative and $j_{l'}(p_{xA}, R_{xA})$ the spherical Bessel function, N the number of the levels included. This is a generalization of the R-matrix approach because we consider reactions with three particles in the exit channel, where the TH-nucleus a in the initial states carries the transferred particle x, which is off-energy-shell.

Eq.3 has four important consequences:

- $\mathbf{A}_{\nu\tau}$ is the same level matrix as in the conventional R-matrix theory [21]. Therefore, it depends on the entry and exit channels reduced width amplitudes γ, energy levels and energy shifts. All of them can be extracted by fitting the experimental THM cross section and then can be used to deduce the $A(x,b)B$ astrophysical factor. In this way, we have an exact parameterisation of the astrophysical factor with no need of extrapolation. HOES effects can affect the phases determining interference and the relative heights of the resonances, but the reduced widths γ, containing the nuclear structure effects, appear in the same way in THM and direct data.
- The presence of the factor $P_{l'}^{-1/2}(k_{xA}, R_{xA})$ eliminates the penetration of Coulomb and centrifugal barriers in the $x+A$ channel, which is the entry channel of the reaction of astrophysical importance. The compensation of this penetrability factor is the main advantage of the THM, since it allows one to measure the astrophysical factor of the reaction of astrophysical interest down to zero energy. Moreover, resonances that can be populated with large l only are not suppressed by the centrifugal barrier, thus they can be observed even in those cases they are very weak in direct measurements, making the THM a powerful spectroscopic tool.
- The use of the PWA implies that normalisation is accomplished by extending the indirect measurement to an energy region where directly measured data are available and scaling the deduced $\gamma-$widths to match the values in the literature. This is because in PWA absolute values are greatly overestimated. However, Eq.3 can be modified as the $a-A$ and the $s-F$ interactions can be treated within the more advanced distorted waves (DWBA) and the CDCC formalisms [14]. This is a very important point as it opens the possibility to make unnecessary normalisation to direct data, at present a major drawback of THM. This is especially important in the investigation of reactions induced by radioactive ion beams, where direct data might be absent or scarce.
- In a THM experiments three particles are emitted in the exit channel. From the measurement of the energies and the angles of emissions of two out of three emitted

IOP Conf. Series: Journal of Physics: Conf. Series **876** (2017) 012013 doi:10.1088/1742-6596/876/1/012013

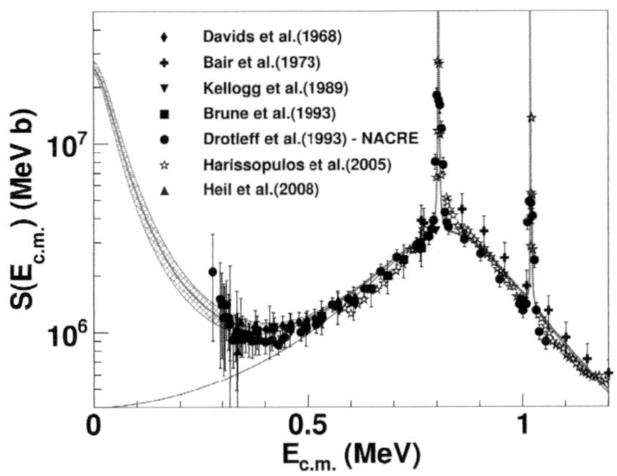

Figure 2. R-matrix calculated S(E)-factor of the $^{13}C(\alpha, n)^{16}O$ reaction (red band) from Ref. [23]. The R-matrix S(E)-factor not including the threshold resonance is displayed by the blue line. Black symbols are used for direct data normalized as in [36]. Different marks are used for each data set, as specified in the inset. See Ref. [23] for more details.

particles, all the kinematic variables can be calculated. The most important parameter for astrophysical applications is the $x - A$ relative energy. Following [7], under the non essential hypothesis that the nucleus a undergoing breakup is at rest in the laboratory system (similar formula can be found for breakup of the projectile), the $x - A$ relative energy can be written as:

$$E_{x-A} = \frac{m_x}{m_x + m_A} E_A - \frac{p_s^2}{2\mu_{sF}} + \frac{\mathbf{p}_s \cdot \mathbf{p}_A}{m_x + m_A} - \varepsilon_{sx} \, , \tag{7}$$

where m_i, $\mathbf{p}_i$ and E_i are the mass, momentum and energy of the i-th particle, μ_{sF} the $s - F$ reduced mass and ε_{sx} the $x - s$ binding energy. Part of the projectile energy is spent to break the impinging nucleus a and thanks to the $x - s$ inter cluster motion, astrophysical energies can be achieved in the $x - A$ channel of the TH reaction using beam energies of few tens of MeV, bypassing Coulomb barrier. Moreover, since high beam energies are used, the electron screening enhancement does not occur. Furthermore, negative E_{x-A} energies can be explored by choosing a suitable combination of beam energy, spectator momentum and target nucleus a. From the THM measurement of sub-threshold states, yielding the reduced widths γ, the ANC can be deduced, making it possible to connect the two indirect approaches [22, 23, 24].

In the past years, a number of studies have been accomplished to validate the THM approach, making the method very robust. For instance, the effect of momentum distribution variations on the deduced astrophysical factors was considered. The use of realistic distributions in the place of the simple Hulthén function in momentum space (in the case of deuteron) has been investigated in [25], while the influence of the experimental momentum distribution in the place of the theoretical one was evaluated in [26]. In both cases, changes comparable or lower than the statistical uncertainty were retrieved, since events in a momentum region of particle s below about 40 MeV/c are usually considered in the data analysis [27]. The use of DWBA in the place of the PWA was tested in [28], showing again significant changes in the deduced astrophysical factors. The method has been extended to measurements of neutron induced reactions [29] and to reactions involving radioactive nuclei [30, 31]

3. The sub-threshold resonance case: the $^{13}C(\alpha, n)^{16}O$ reaction

In a class of evolved stars, those belonging to the so-called asymptotic giant branch (AGB), very peculiar conditions are present making it possible to synthesise nuclei heavier than iron through a succession of slow neutron captures, followed by decay of the formed unstable nuclei. This path to heavy nuclei is named s-process and it is responsible of the production of $\sim 50\%$ of nuclei with A $\gtrsim 56$. In AGB stars, protons from the outer layers are mixed downward and quickly captured by carbon nuclei, eventually leading to the formation of a ^{13}C pocket [33]. Then, ^{13}C nuclei emits neutrons through the $^{13}\mathrm{C}(\alpha,n)^{16}\mathrm{O}$ reaction, at temperatures varying between 0.8×10^8 K and 1×10^8 K [34], which can be captured by seed nuclei to build up heavier nuclei, later transported to the stellar surface.

At $0.9\,10^8$ K, the energy range where the $^{13}\mathrm{C}(\alpha,n)^{16}\mathrm{O}$ reaction is most effective, the Gamow window [2], is $\sim 140-230$ keV. In such region, its direct measurement is very challenging because of the Coulomb barrier, exponentially suppressing the cross section, and the interplay between a threshold resonance determined by the population of the 6.356 MeV level in ^{17}O and atomic electron screening. Indeed, at ~ 300 keV the cross section of the $^{13}\mathrm{C}(\alpha,n)^{16}\mathrm{O}$ reaction is already as low as $\sim 10^{-10}$ b, thus direct measurements stopped at ~ 280 keV [35]. Moreover, direct measurements show contradicting results also at MeV energies, owing to uncertainties in the absolute normalisation. Therefore, extrapolation is necessary at present to assess the astrophysical factor of the $^{13}\mathrm{C}(\alpha,n)^{16}\mathrm{O}$ reaction at astrophysical energy, but large errors might be introduced owing to the systematic uncertainty affecting high energy data. Therefore, [36] employed a broad data set including renormalised $^{13}\mathrm{C}(\alpha,n)^{16}\mathrm{O}$ astrophysical factors to perform an extensive R-matrix fit; a 140 keV astrophysical factor $\mathrm{S}(140\,\mathrm{keV}) = 2.2^{+1.1}_{-0.8}\times 10^6$ MeVb was then obtained. We consider 140 keV as this is the smallest energy of interest for astrophysics, the hardest to reach by conventional approaches and the most subject to systematic effects in the case of extrapolation. The renormalised data-set is shown in Fig.2 as black symbols.

The THM S-factor is shown instead as a red band in Fig.2. Thanks to our approach, a very accurate result is obtained with no need of extrapolation down to zero energy, making it possible to supply a very accurate reaction rate for astrophysical modelling. This is possible since we could directly access the 6.356 MeV level in ^{17}O, sitting at -3 keV in the $^{13}\mathrm{C}-\alpha$ relative energy spectrum. The recommended THM S-factor at $\mathrm{E}_{c.m.} = 140$ keV is $3.2\pm 0.5\times 10^6$ MeVb, about 45% larger than the value provided by [36]. The most important result is a significant reduction of the uncertainty on the $^{13}\mathrm{C}(\alpha,n)^{16}\mathrm{O}$ S-factor at the Gamow peak, which decreased from about 43% to about 16%. Recently, two works [37, 38] have determined new interesting spectroscopic information about the 6.356 MeV. The consequences on the $^{13}\mathrm{C}(\alpha,n)^{16}\mathrm{O}$ S-factor are presently under investigation [39].

4. Resonances above the threshold: the $^{19}\mathrm{F}(p,\alpha_0)^{16}\mathrm{O}$ reaction

Since the s-process is quite complicated, involving large nuclear physics input as well as peculiar astrophysical conditions, the possibility to constrain such boundary conditions is very appealing. Fluorine might represent a strong constraint of stellar internal structure since its abundance is very sensitive to the physical conditions in the inner layers of AGB stars [40]. This entails the understanding of fluorine nucleosynthesis, which is incomplete to date. A possible explanation is the poor accuracy of our present picture of the fluorine destruction due to extra-mixing processes [40], where fluorine is exposed to protons at temperatures $\lesssim 4\times 10^7$ K.

The $^{19}\mathrm{F}(p,\alpha)^{16}\mathrm{O}$ reaction is the main destruction channel of fluorine in this scenario. However, only one set of direct data is available so far at the energies where fluorine burning is most effective ($\mathrm{E}_{c.m.} \leq 300$ keV, the Gamow energy [2]), still with quite large errors [41]. Furthermore, only the α_0 channel, corresponding to the emission of α-particles off ^{20}Ne compound system leaving ^{16}O in its ground state, has been investigated, being considered the larger contributor to the total cross section [41]. Before, only extrapolations were available [42] below about 500 keV, showing a non resonant behaviour, sharply contradicting the trend of the astrophysical factor at higher energies. This very simple recommended extrapolation to astrophysical energies has triggered the reassessment of the nuclear reaction rates involved in

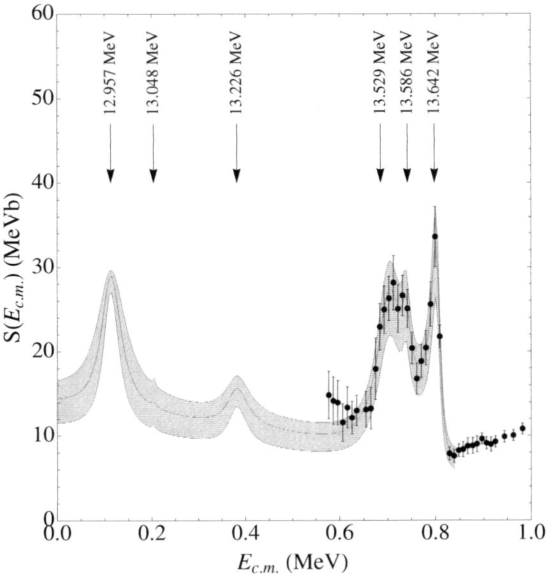

Figure 3. THM astrophysical factor of the $^{19}\mathrm{F}(p,\alpha_0)^{16}\mathrm{O}$ reaction, normalized to the data from [45] above 600 keV, as shown in [44]. The middle red line marks the recommended S-factor, while the upper and lower red lines stand for upper and lower limits set by combined statistical, normalization and energy shift error. The solid symbols represent the direct astrophysical factor in [45]. Finally, the arrows mark the ^{20}Ne states contributing to the S(E)-factor.

fluorine production and destruction, in particular by using the THM in its version modified to deal with resonant reactions [43, 44].

To this purpose, the QF $^2\mathrm{H}(^{19}\mathrm{F},\alpha^{16}\mathrm{O})n$ reaction at 50 MeV beam energy was measured by means of a ^{19}F beam impinging onto deuterated polyethylene targets, thus using deuterons to transfer protons and induce the $^{19}\mathrm{F}(p,\alpha)^{16}\mathrm{O}$ QF reaction. More details on the experiment are given in [43]. Here we underscore that by using Eq.3 we derived the reduced widths of a number of resonances and, in particular, of a 113 keV peak sitting right inside the Gamow window, which was later confirmed by [41] and might have important consequences for astrophysics. The $p-{}^{19}\mathrm{F}$ relative energy spectrum spanned an energy interval from 0 to about 1 MeV, making it possible to normalise the THM astrophysical factor to the existing direct data. In the original work [43], THM data were normalised to a weighed average of direct data in the energy window $0.6-0.8$ MeV [42]. Later, new direct data were made available in the normalisation energy region [45], leading to a reanalysis of the THM data [44].

Fig. 3 shows the S(E)-factor calculated with the resonance parameters from the fitting of THM data below 600 keV. The middle red curve marks the S(E)-factor computed using the parameters from the best fit, while the red band arises from the uncertainties on the resonance parameters, due to the combined statistical and systematic error. An average error of 20% is obtained. At present, the main source of uncertainty is due to the non resonant contribution to the astrophysical factor, since the one given in [42] is based on a very simple calculation. New direct measurements are of utmost importance to have a more realistic non-resonant contribution at low energies.

A new THM experiment has been recently performed to improve the energy resolution affecting THM data, which is a second important source of indetermination on the THM S-factor. Indeed, owing to the occurrence of many resonances below 600 keV, suitable energy resolution is at order to disentangle the occurring resonances. The poor energy resolution in the original experiment [43] prevented us to correctly identify resonances around 200 keV, where a small peak is apparent in Fig.3. Direct data [41] showed a peculiar behaviour at the lowest energies, demonstrating the occurrence of a broad 2^+ state at 251 keV, which was misidentified in the pioneering work [43] owing to the interplay between the poor energy resolution and its width (162 keV). Therefore, a direct comparison between the THM data [43, 44] and those in

Ref.[41] is presently impossible. This last work has triggered the new improved THM experiment mentioned above, aimed at passing the resolution of the pioneering work [43]. Preliminary results suggest an agreement between direct and indirect data and the manuscript is in preparation [46].

References

[1] Rolfs C E and Rodney W S 1998 *Cauldrons in the cosmos* (Chicago: University of Chicago Press)
[2] Iliadis C 2007 *Nuclear Physics of Stars* (New York: Wiley)
[3] Best A et al. 2016 *Eur. Phys. J.* A **52** 72
[4] Lamia L , Spitaleri c, Pizzone RG et al. 2013 *The Astrophysical Journal* **768** 65
[5] Tumino A, Sparta R, Spitaleri C et al. 2014 *The Astrophysical Journal* **785** 96
[6] Strieder F, Rolfs C, Spitaleri C and Corvisiero P . 2001 *Naturwissenschaften* **88** 461
[7] Tribble R E, Bertulani CA, La Cognata M et al. 2014 *Reports on Progress in Physics* **77** 106901
[8] Mukhamedzhanov A M, Clark HL, Gagliardi CA et al. 1997 *Phys. Rev.* C **56** 1302
[9] Baur G and Rebel H 1994 *Journal of Physics G Nuclear Physics* **20** 1
[10] Spitaleri C, Mukhamedzhanov A M, Blokhintsev, L. D et al. 2011 *Physics of Atomic Nuclei* **74** 1725
[11] Spitaleri C 1991 *Proceedings of the 5Th Winter School on Hadronic Physics: Problems of Fundamental Modern Physics II* (Singapore: World Scientific) 21
[12] Baur G 1986 *Physics Letters* B **178** 135
[13] Mukhamedzhanov A M, Blokhintsev LD, Irgaziev BF et al. 2008 *Journal of Physics G Nuclear Physics* **35** 014016
[14] Mukhamedzhanov A M 2011 *Phys. Rev.* C **84** 044616
[15] La Cognata M, Romano S, Spitaleri C et al. 2006 *European Physical Journal* A **27** 249
[16] La Cognata M, Spitaleri C, Tribble R E et al. 2008 *Journal of Physics G Nuclear Physics* **35** 014014
[17] La Cognata M, Goldberg VZ, Mukhamedzhanov AM et al. 2009 *Phys. Rev.* C **80** 012801
[18] La Cognata M , Spitaleri C, Mukhamedzhanov A M et al. 2010 *The Astrophysical Journal* **723** 1512
[19] Palmerini S, La Cognata M, Cristallo S. et al. 2011 *The Astrophysical Journal* **729** 3
[20] Palmerini S, Sergi, M. L., La Cognata, M et al. 2013 *The Astrophysical Journal* **764** 128
[21] Lane A M and Thomas R G 1958 *Reviews of Modern Physics* **30** 257
[22] La Cognata M , Spitaleri C, Trippella O et al. 2012 *Physical Review Letters* **109** 232701
[23] La Cognata M, Spitaleri C, Trippella O et al. 2013 *The Astrophysical Journal* **777** 143
[24] La Cognata M , Spitaleri C, Cherubini S et al. 2016 *Acta Physics Polonica* B **47** 681
[25] Lamia L, La Cognata M, Spitaleri C et al. 2012 *Phys. Rev.* C **85** 025805
[26] Pizzone R G, Spitaleri C, Cherubini S et al. 2005 *Phys. Rev.* C **71** 058801
[27] Spitaleri C, La Cognata M, Lamia L et al. 2016 *European Physical Journal* A **52** 77
[28] La Cognata M, Spitaleri C, Mukhamedzhanov A. et al. 2010 *Nucl. Phys.* A **834** 658C
[29] Gulino M, Spitaleri C, Tang X D et al. 2013 *Phys. Rev.* C **87** 012801
[30] Cherubini C, Gulino M, Spitaleri C et al. 2015 *Phys. Rev.* C **92** 015805
[31] Pizzone R G, Roeder B T, McCleskey M et al. 2016 *European Physical Journal* A **52** 24
[32] Heil M, Detwiler R, Azuma R E et al. 2008 *Phys. Rev.* C **78** 025803
[33] Gallino R, Arlandini C, Busso M; et al. 1998 *Astrophys. J.* **497** 388
[34] Busso M, Gallino R, Wasserburg G J et al. 1999 *Annual Review of Astronomy and Astrophysics* **37** 239
[35] Drotleff H W, Denker H, Knee H et al. 1993 *Astrophys. J.* **414** 735
[36] Heil M, Kaeppeler F, Uberseder E et al. 2008 *Phys. Rev.* C **78** 025803
[37] Avila M L, Rogachev G V, Koshchiy E et al. 2015 *Phys. Rev.* C **91** 048801
[38] Faestermann T, Mohr P, Hertenberger R et al. 2015 *Phys. Rev.* C **92** 052802
[39] Trippella O and La Cognata M 2017 *Astrophys. J.* **41** 837
[40] Lugaro M, Ugalde C, Karakas AI et al. 2004 *Astrophys. J.* **615** 934
[41] Lombardo I, Dell'Aquila D, Di Leva A et al. 2015 *Phys. Lett.* B **748** 178
[42] Angulo C, Arnould M, Rayet M et al. 1999 *Nuclear Physics* A **656** 3
[43] La Cognata M, Mukhamedzhanov A M, Spitaleri C et al. 2011 *The Astrophysical Journal Letters* **739** L54
[44] La Cognata M, Palmerini S, Spitaleri C et al. 2015 *The Astrophysical Journal* **805** 128
[45] Lombardo I, Dell'Aquila D, Campajola L et al. 2013 *Journal of Physics G Nuclear Physics* **40** 125102
[46] Indelicato I, La Cognata M, Spitaleri C et al. 2017 *Astrophys. J.*, in preparation

IOP Conf. Series: Journal of Physics: Conf. Series **876** (2017) 012014 doi:10.1088/1742-6596/876/1/012014

Confinement in a correlated Instanton-Dyon Ensemble

Miguel Angel Lopez-Ruiz[1], **Yin Jiang**[1,2], **Jinfeng Liao**[1]
[1]Physics Department and Center for Exploration of Energy and Matter, Indiana University, 2401 N Milo B. Sampson Lane, Bloomington, IN 47408, USA
[2]Institut fuer Theoretische Physik, Philosophenweg 16, D-69120 Heidelberg, Germany

E-mail: malopezr@indiana.edu

Abstract. Confinement is a remarkable nonperturbative phenomena emerging from QCD and QCD-like theories. A theoretical understanding of these transitions and their interrelations is of fundamental importance. While it is widely perceived that their dynamics arises from nontrivial topological configurations in Yang-Mills theories, a concrete and sophisticated realization of such idea is an outstanding challenge. We report significant progress along this direction by the construction of a new framework based on correlated ensemble of instanton-dyons, namely the constituents of the finite-temperature instantons with non-trivial holonomy. We present a comprehensive numerical study of confinement properties in SU(2) Yang-Mills theory at finite temperature, obtaining important observables such as the effective holonomy potential, the static quark potentials from Polyakov loop correlators as well as spatial Wilson loops, among others.

1. Introduction

The Quantum Chromodynamics, or QCD, is established as the fundamental quantum field theory of strong nuclear force underlying all of nuclear physics. Despite its great success in describing an impressive variety of nuclear phenomena in Nature, a key aspect of QCD remains mysterious and poses a great challenge to our understanding. While the theory has quarks and gluons as its basic degrees of freedom in the Lagrangian, the colored quarks and gluons are absent from the observed physical spectrum in which the various color-singlet hadronic states emerge instead. This phenomenon, often referred to with the broad term "confinement", occurs also in a wide variety of QCD-like theories and notably in pure Yang-Mills theories. The latter fact makes it obvious that confinement arises from the nonperturbative gauge dynamics in the gluonic sector.

At finite temperature, the expectation value of the Polyakov loop at spatial infinity (holonomy)

$$L_\infty \equiv \lim_{|\vec{x}|\to\infty} \frac{1}{N_c} \mathrm{Tr}\, \mathcal{P} \exp\left(i \int_0^{1/T} \mathrm{d}\tau A_4 \right) \tag{1}$$

plays the role of the order parameter in the deconfinement phase transition; suggesting that topological gauge configurations with non-trivial holonomy may drive the mechanism of confinement.

Based on the *caloron solutions with nontrivial holonomy and nontrivial topology* to the classical Yang-Mills equations, known as the KvBLL calorons [1, 2, 3]; in particular, its "constituent" dyon fields (also called *instanton-dyons*), a promising approach has emerged for understanding the nonperturbative phenomenon of confinement through a statistical ensemble of such objects.

This talk is based on the extensive work presented in [4] where we report a thorough numerical investigation of the confinement dynamics in SU(2) Yang-Mills theory by constructing a statistical ensemble of correlated instanton-dyons. Here, we summarize our high precision results for the holonomy potential, the order parameter for confinement transition, as well as the temporal and spatial Wilson loops with an emphasis on the influence of finite volume effects.

2. Construction of Correlated Instanton-Dyon Ensemble

2.1. The Partition Function

The partition function of the instanton-dyon/antidyon ensemble is constructed from the one-loop quantum weight of the KvBLL caloron in the limit of large dyon separation [5], then extended to arbitrary number dyons and antidyons, resulting in

$$\mathcal{Z} = e^{-VP(\nu)} \sum_{\substack{N_M, N_L,\\ N_{\bar{L}}, N_{\bar{M}}}} \frac{1}{N_L! N_M! N_{\bar{L}}! N_{\bar{M}}!} \int \prod_{l=1}^{N_L} f_L T^3\, \mathrm{d}^3 r_{L_l} \prod_{m=1}^{N_M} f_M T^3\, \mathrm{d}^3 r_{M_m}$$
$$\times \prod_{\bar{l}=1}^{N_{\bar{L}}} f_{\bar{L}} T^3\, \mathrm{d}^3 r_{\bar{L}_{\bar{l}}} \prod_{\bar{m}=1}^{N_{\bar{M}}} f_{\bar{M}} T^3\, \mathrm{d}^3 r_{\bar{M}_{\bar{m}}} \det(G_D) \det(G_{\bar{D}})\, e^{-V_{D\bar{D}}}, \tag{2}$$

where $f_M = \Gamma S^2 e^{-\nu S} \nu^{\frac{8\nu}{3}-1}$ and $f_L = \Gamma S^2 e^{-\bar{\nu} S} \bar{\nu}^{\frac{8\bar{\nu}}{3}-1}$ are the fugacities per (anti)dyon species L, M, $\bar{L}$, $\bar{M}$, and the holonomy parameter $\nu(\bar{\nu} = 1-\nu)$ is defined through $L_\infty = \cos(\pi\nu)$. The sum runs over the total number of dyons and antidyons in the system and the measure is given by the product of an uncorrelated part $\det(G_D)\det(G_{\bar{D}})$, namely the corresponding *"Diakonov determinants"* [6], and a correlated factor $e^{-V_{D\bar{D}}}$ which introduces dyon-antidyon interactions given by [7][8]

$$V_{D\bar{D}} = \begin{cases} \sum_{j>\bar{j}} 2S\left(\frac{1}{\zeta_j} - 1.632\, e^{-0.704\zeta_j}\right) e^{-M_D r_{j\bar{j}}} & \text{if } \zeta_j > \zeta_j^0, \text{ for } L\bar{L}, M\bar{M} \\ \sum_{i>j} V_{ij}^C & \text{if } \zeta_j < \zeta_j^0, \text{ for } \begin{matrix} LL, \bar{L}\bar{L}, MM, \\ \bar{M}\bar{M}, L\bar{L}, M\bar{M} \end{matrix} \\ \sum_{i,\bar{j}} \frac{S}{\pi T r_{i\bar{j}}} e^{-M_D r_{i\bar{j}}} & \text{for } \bar{M}L, \bar{L}M \\ 0 & \text{for } LM, \bar{L}\bar{M}, \end{cases} \tag{3}$$

where $\zeta_j = 2\pi\nu_j T r_{j\bar{j}}$, M_D is a screening mass introduced as a parameter, $V_{j\bar{j}}^C = \frac{\nu_j V_c}{1+e^{(\zeta_j - \zeta_j^0)}}$ is a repulsive core potential where V_c and ζ_j^0 are also parameters which need to be tuned for the simulations.

For simplicity, all quantities are rescaled with temperature T; however, the temperature dependence is defined through the running of the coupling constant in the caloron action S, which at first loop approximation, is given by

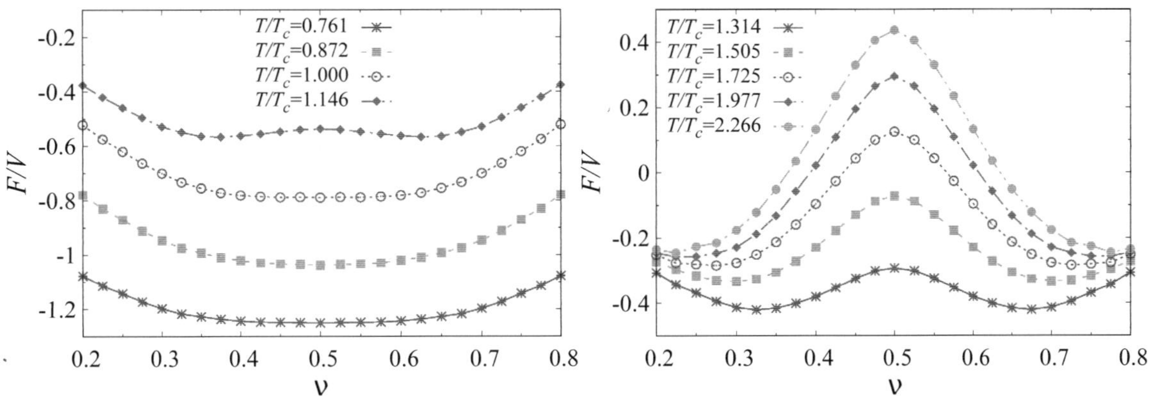

Figure 1. Free energy density $F = -\log \mathcal{Z}$ of the dyon ensemble in the density range $0 \leq n_D \leq 0.5$.

$$S(T) = \frac{8\pi^2}{g^2(T)} = \frac{22}{3} \log\left(\frac{T}{\Lambda}\right), \tag{4}$$

where Λ is the scale parameter in the regularization. Therefore, by varying S as a parameter in the simulation, one can define Λ at the observed critical point ($T = T_c$) and then establish the temperature dependence of the observables as T/T_c.

Through Monte Carlo methods, the ensemble is simulated on a box with periodic boundary conditions of volume $V_0 = 43.37$ and a total number of (anti)dyons of each kind running from $N_D = 0$ to 22, i.e. the dyon density range spanned was $n_D = 0$ to 0.5.

3. Confinement-Deconfinement Transition

3.1. The Holonomy Potential

Above some critical temperature T_c, $SU(2)$ pure gauge theory is expected to have a second order deconfinement phase transition, with $\langle L_\infty \rangle$ as the order parameter characterizing its criticality. When the gauge field is that of the KvBLL caloron, the order parameter takes the form $\langle L_\infty \rangle = \cos(\pi\nu)$, so one expects to observe the breaking of the Z_2 symmetry of the theory through the free energy density, either as a function of L_∞ or ν.

1 shows the free energy density for $S = 5, 6, \ldots, 13$. It was found that for $5 \leq S \leq 7$ the minimum of the free energy density lies at $\nu_{\rm min} = 0.5$, namely maximal non-trivial holonomy. For $S > 7$ the shape of F/V becomes that of a symmetric double well potential with $\nu_{\rm min} < 0.5$. Therefore, the critical temperature is defined at $S = 7$, fixing the scale parameter at $\Lambda = 0.385 T_c$.

Universality and the Svetitsky-Yaffe conjecture [9], allow to explore the role of $\langle L_\infty \rangle$ as an order parameter of the deconfinement phase transition, given that it should follow the power law

$$\langle L_\infty \rangle = a(T/T_c - 1)^\beta \left[1 + b(T/T_c - 1)^\omega\right], \tag{5}$$

for temperatures close to T_c. 2 shows a fit from the numerical results of the dyon ensemble in the interval $1 \leq T/T_c \leq 1.505$. This result shows the continuous nature of the phase transition and is in qualitative agreement with the lattice results from [10].

As for the expectation value of dyon densities, depicted in 3, one can see that at $T \leq T_c$, the L and M type densities are fairly equal. In the confined phase, the preferred holonomy

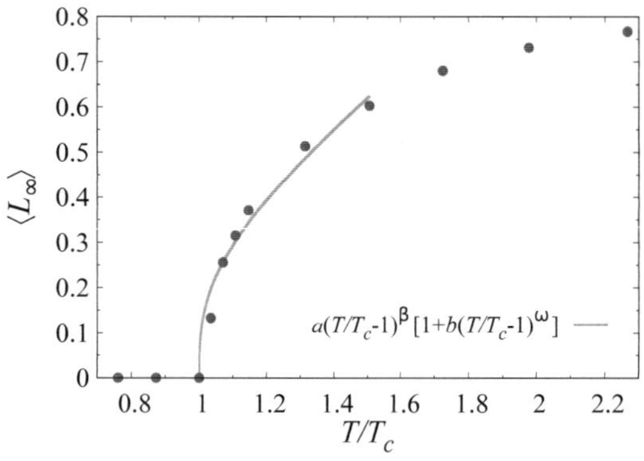

Figure 2. L_∞ as an order parameter of the phase transition. Fitted to the critical exponents β and ω of the 3D Ising model.

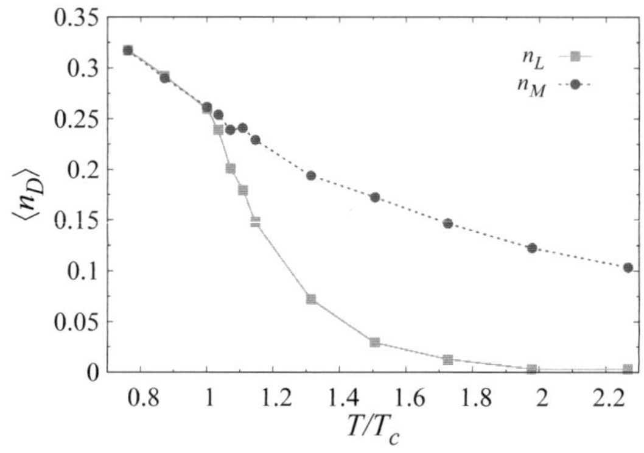

Figure 3. Temperature dependence of the ensemble average of dyon densities.

corresponds to the maximal non-trivial one where both dyon types have the same core radius and therefore equal weight in the partition function. For $T > T_c$, the prefered holonomy starts to shift towards the trivial one ($\nu \to 0$) and the M dyons become larger and larger.

3.2. The Polyakov Loop Correlator and the Spatial Wilson Loop

In the study of confinement, it is of most importance to look at the interaction between a static (infinitely heavy) quark-antiquark pair. A successful confining theory should have an asymptotically linear rising potential at large separations between the static sources in the form $F_{q\bar{q}}|_{|\vec{x}-\vec{y}|\to\infty} \approx \sigma_e|\vec{x}-\vec{y}|$, where σ_e is the so called *electric string tension.* At finite temperature, the color averaged heavy quark-antiquark potential $F^{\text{avg}}_{q\bar{q}}$ is defined through the expectation value of traced Polyakov loop correlators. Moreover, according to the color decomposition $2\otimes\bar{2} = 1\oplus 3$, an $SU(2)$ quark-antiquark pair can interact through a singlet and a triplet channel like

$$e^{F^{\text{avg}}_{q\bar{q}}} = \frac{1}{4}e^{-F^{1}_{q\bar{q}}} + \frac{3}{4}e^{-F^{3}_{q\bar{q}}}, \tag{6}$$

where

$$e^{-F^{\text{avg}}_{q\bar{q}}} \equiv \frac{1}{4}\left\langle \text{Tr}\ L^\dagger(\vec{x})\ \text{Tr}\ L(\vec{y})\right\rangle \quad \text{and} \quad e^{-F^{1}_{q\bar{q}}} \equiv \frac{1}{2}\left\langle \text{Tr}\left[L^\dagger(\vec{x})L(\vec{y})\right]\right\rangle. \tag{7}$$

On 4 we show the singlet channel free energy $F^1_{q\bar{q}}$ as a function of interquark separation $|\vec{x}-\vec{y}|$ for different temperatures. One can see, that below T_c, there is a linear rising potential with almost constant slope σ_e; however, for $T > T_c$, the slope starts to decrease until it reaches values close to zero, implying a deconfined phase.

As for the spatial Wilson loop, it is known that at finite temperature it does not provide a good measure of the deconfinement phase transition, given that even above T_c it shows area law behavior and an increase of the *magnetic string tension* σ_m [11, 12, 13, 14]; nonetheless, the restoration of Lorentz symmetry (Euclidean $O(4)$) at $T \to 0$ suggests that in this limit, σ_m should coincide with the electric string tension σ_e. Therefore, the interest is to show that the spatial Wilson loop defined as

$$W_C \equiv \frac{1}{2}\,\text{Tr}\ \mathcal{P}\exp\left[i\oint_C \mathrm{d}x_i A_i(x)\right], \tag{8}$$

follows the area law

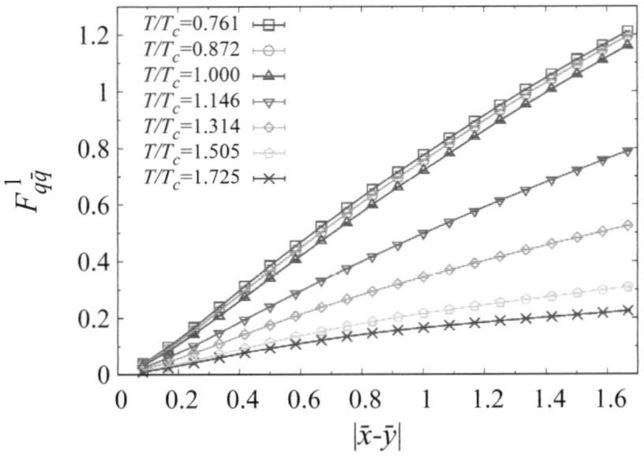

Figure 4. Singlet channel of the static quark-antiquark potential in the fundamental representation at confined and deconfined phases for $\nu = 0.5, 0.5, 0.35$ and 0.275, in order of increasing temperature.

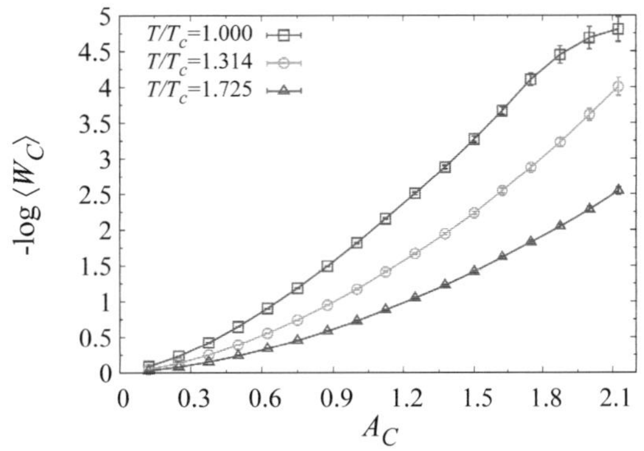

Figure 5. Area law for the spatial Wilson loop in the fundamental representation at confined and deconfined phases.

$$\langle W_C \rangle \sim e^{-\sigma_m A_C} \tag{9}$$

in both confining and deconfined phases. On 5, the negative logarithm of $\langle W_C \rangle$ in the fundamental representation is plotted as a function of A_C and it can be seen that for large contour areas, indeed has an almost linear rising.

To end this section, we recall that in the units used in this work, the string tension is dimensionless, meaning that when restoring physical units it goes as $\sigma_m \rightarrow \sigma_m/T^2$. As has been established before, σ_m increases with T; however, σ_m/T^2 should decrease as the temperature rises [13], which is in fact in agreement with our results.

4. Finite volume effects

To look into the possible effects which a finite volume have on the computed observables, we performed tests increasing the volume of the box to two and three times the previous volume used V_0.

1 contains numerical values of ensemble averages of the dyon densities at different temperatures as well as the free energy density for $\nu_{\min}$ obtained at V_0. It can be seen that the volume effects in the densities are significantly smaller than in the free energy density, which in fact are considerably small.

Another test on the influence of volume was done with the static quark-antiquark potentials. On 6 we show the results of the singlet channel potential calculated in a box twice the size of the previous volume used. The curved tails which were present at large distances, seemed to appear only at the edge of the box, suggesting it is merely a finite volume effect of the simulation. At intermediate distances both potentials match substantially well showing the expected linear rising behavior.

5. Conclusion

Confinement is a remarkable nonperturbative phenomenon in pure Yang-Mills and QCD-like theories. The mechanism of confinement remains a significant challenge to our understanding and is generally believed to pertain to certain nontrivial topological configurations of the gluonic

Table 1. Volume dependence of the free energy density and ensemble averages of dyon densities for $V = V_0, 2V_0$ and $3V_0$, with $V_0 = 43.37$.

	T/T_c	V_0	$2V_0$	$3V_0$
	0.761	-1.251	-1.275	-1.283
F/V	1.000	-0.790	-0.813	-0.826
	1.314	-0.416	-0.433	-0.444
	0.761	0.317	0.320	0.315
$\langle n_L \rangle$	1.000	0.259	0.259	0.260
	1.314	0.075	0.077	0.077
	0.761	0.317	0.315	0.312
$\langle n_M \rangle$	1.000	0.261	0.262	0.259
	1.314	0.215	0.216	0.215

sector. The recently found KvBLL caloron solutions with nontrivial holonomy, consisting of constituent dyons, have provided a concrete and promising path of investigation. We conclude that an ensemble of such objects correctly produces the various essential features of the confinement dynamics from above to below the transition temperature. These features include the evolution of holonomy potential with temperature, a second order phase transition in terms of the order parameter (Polyakov loop expectation value), the linear static quark-anti-quark potential, etc. Given such success, it appears reasonable to believe that the ensemble of correlated instaton-dyons may indeed hold the key of confinement mechanism.

Acknowledgements
The authors are particularly grateful to E. Shuryak for many helpful discussions. The authors also thank R. Larsen, E.-M. Ilgenfritz, M. Faber, R. Pisarski, I. Zahed, and A. Zhitnitsky for useful discussions and communications. This work is supported by the National Science Foundation under Grant No. PHY-1352368. MALR is in addition supported by CONACyT under Doctoral supports Grants No. 669645. The computation of this research was performed on IU's Big Red II cluster that is supported in part by Lilly Endowment, Inc. (through its

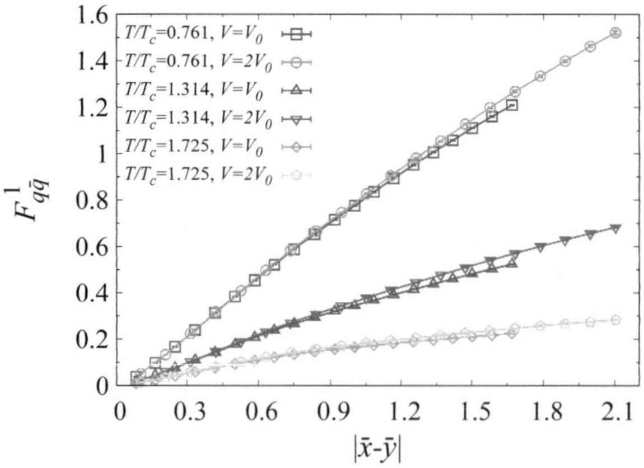

Figure 6. Volume effects on the singlet channel of the static quark-antiquark potential at confined and deconfined phases.

support for the Indiana University Pervasive Technology Institute) and in part by the Indiana METACyt Initiative.

References

[1] T. C. Kraan and P. van Baal, *Nucl. Phys.* B **533**, 627 (1998) [hep-th/9805168].
[2] T. C. Kraan and P. van Baal, *Phys. Lett.* B **435**, 389 (1998) [hep-th/9806034].
[3] K. M. Lee and C. h. Lu, *Phys. Rev.* D **58**, 025011 (1998) [hep-th/9802108].
[4] M. A. Lopez-Ruiz, Y. Jiang and J. Liao, *Preprint* arXiv:1611.02539 [hep-ph].
[5] D. Diakonov, N. Gromov, V. Petrov and S. Slizovskiy, *Phys. Rev.* D **70**, 036003 (2004) [arXiv:hep-th/0404042].
[6] D. Diakonov and V. Petrov, *Phys. Rev.* D **76**, 056001 (2007) [arXiv:0704.3181 [hep-th]].
[7] R. Larsen and E. Shuryak, *Nucl. Phys.* A **950**, 110 (2016) [arXiv:1408.6563 [hep-ph]].
[8] R. Larsen and E. Shuryak, *Phys. Rev.* D **92**, no. 9, 094022 (2015) [arXiv:1504.03341 [hep-ph]].
[9] B. Svetitsky and L. G. Yaffe, *Nucl. Phys.* B **210**, 423 (1982).
[10] S. Digal, S. Fortunato and P. Petreczky, *Phys. Rev.* D **68**, 034008 (2003) [hep-lat/0304017].
[11] C. Borgs, *Nucl. Phys.* B **261**, 455 (1985).
[12] E. Manousakis and J. Polonyi, *Phys. Rev. Lett.* **58**, 847 (1987).
[13] G. S. Bali, J. Fingberg, U. M. Heller, F. Karsch and K. Schilling, *Phys. Rev. Lett.* **71**, 3059 (1993) [hep-lat/9306024].
[14] L. Karkkainen, P. Lacock, D. E. Miller, B. Petersson and T. Reisz, *Phys. Lett.* B **312**, 173 (1993) [hep-lat/9306015].

IOP Conf. Series: Journal of Physics: Conf. Series **876** (2017) 012015 doi:10.1088/1742-6596/876/1/012015

Study of the 12C excited states above the Hoyle State.

E. López-Saavedra[1], L. Acosta[1], V. Araujo[1], F. Favela[1], A. Huerta[1], J. Aspiazu[2], G. Murillo[2], R. Policroniades[2], P. Santa Rita[1], A. Varela[2] and E. Chávez[1]

1. Instituto de Física, Universidad Nacional Autónoma de México. 04510, Ciudad de México, México.

2. Departamento de Aceleradores, Instituto Nacional de Investigaciones Nucleares, Carretera México-Toluca S/N, La Marquesa Ocoyoacac, 52750, Estado de México, México.

E-mail: eilenslopez.154@gmail.com

Abstract. In this work we study the low-lying excited states of ^{12}C, especially above the Hoyle state (0^+, 7,654 MeV) through the use of the $^{14}N(d,\alpha)^{12}C$ reaction. The EN-Tandem at ININ delivered deuteron beams between 2.5 and 7.5 MeV. Typical beam intensities were 20-50 nA. Two different compounds were used to produce thin films: Si_3N_4 (150 nm) and of $C_5H_5N_5$ (10 µm). Angular distributions of emitted α-particles were measured at each energy. The first results of the analysis are presented including quantum number assignments (energy, spin and parity) of the excited states populated.

1. Introduction

The production of the carbón nucleus is a key-step in stellar nucleosynthesis. Its most abundant isotope, ^{12}C, is produced through the triple α-process and the short-lived ^{8}Be ground state as an intermediate state [1].

The so called "Hoyle state", the 7.654 MeV 0^+ in ^{12}C, plays an important role in a variety of problems of nuclear astrophysics like the abundance of the elements in the universe and the stellar nucleosynthesis process as a whole. From the nuclear structure point of view, there are unanswered questions regarding its geometry. Even though the standard shell-model approaches and the advanced no-core calculations reproduce many of the ^{12}C resonances, they fail to reproduce such excited state [2-10].

Recent experiments report on the measurement of a new high spin (5^-) state predicted by the Algebraic Cluster Model (ACM). This model predicts other states unseen up to date. So the experimental observation (or not) of these states is crucial in the understanding of the nuclear structure of ^{12}C [11-13].

The purpose of this work is to study the nuclear structure of ^{12}C by the measurement of the $^{14}N(d,\alpha)^{12}C$ reaction at different energies between 2.5-7.5 MeV, using two targets the Si_3N_4 and the $C_5H_5N_5$, the excited states populated are studied, especially those above the Hoyle state (0^+, 7.654 MeV).

2. Experimental Set-up

Two experiments were performed at the EN-Tandem accelerator of the National Institute of Nuclear Research (ININ), Beams of deuterium between a range of energies of 2.5 and 7.5 MeV were used to bombard two targets: Si_3N_4 (silicon nitride) and $C_5H_5N_5$ (adenine), in order to study the $^{14}N(d,\alpha)^{12}C$

IOP Conf. Series: Journal of Physics: Conf. Series **876** (2017) 012015 doi:10.1088/1742-6596/876/1/012015

reaction. A detector array of 6 passivated implanted planar silicon (PIPS) detectors was used for the detection of the charged particles. The array was arranged in an angular range between 105°-165°. The detectors were calibrated with beam energies of 2.5, 5.0 ,6.5 and 7.0 MeV.The spectra obtained were analysed with the SIMRA software [14].

Angular distributions for each excited state populated in ^{12}C were obtained between the beam energy range of 2.5 and 7.5 MeV with 250 keV intervals for a total of 21 runs.

Figure 1 shows the scattering chamber used in the experiments and the arrangement of the detectors.

Figure 1. Experimental setup used in both experiments, on the left hand side the scattering chamber, on the right hand side, the detector array at backward angles.

3. Data Analysis and Preliminary Results

As mentioned above, the SIMRA software was used for the the identification of the excited states populated. In Figures 2 and 3 are shown two spectra obtained in the experiments with targets of Si_3N_4 and $C_5H_5N_5$ respectively.

Despite the resolution obtained for the thin silicon nitride film (70 nm) for the ground state (α_0), several resonances in ^{29}Si were populated, making ambiguous the identification of the cross section for the reaction of interest.

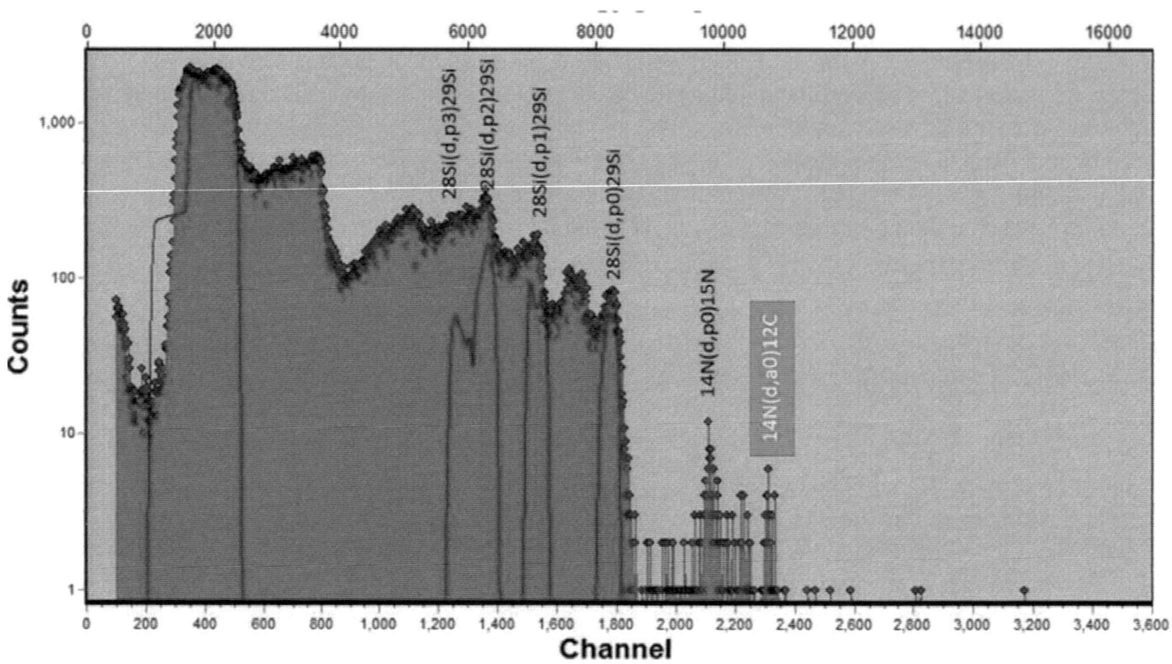

Figure 2. Spectrum obtained in the experiment with a Si_3N_4 target, θ_{lab}: 135 ° E_{lab}: 3 MeV.

Because the poor statistics and the presence of other reactions a second target was required, a $C_5H_5N_5$ target (426μg/cm2), much thicker than the Si_3N_4 film. An example of the spectra obtained in this experiment is shown in Figure 3. Here it was possible to identify several excited states of ^{12}C from the reaction of interest. In addition, we found other reactions that sometimes overlap with those of our interest, these reactions are shown in Figure 3.

In this experiment, the resolution was worse than that in the experiment with Si_3N_4 target, but the statistic were higher. In this experiment we were able to measure the cross sections corresponding to the ground state (0.0 MeV) and the first seven excited states of ^{12}C (4.43 MeV, 7.654 MeV, 9.641 MeV, 10.3 MeV, 10.84 MeV, 11.16 MeV, 11.8 MeV)

In this preliminary work we present experimental angular distributions for the states in ^{12}C with excitation energies of 0.0 MeV ,4.43 MeV, 7.654 MeV and 9.641 MeV respectively, at E_{lab} = 4.25 MeV. A theoretical DWBA calculation (using the DWUCK 4 code [15]) is also shown for each case.

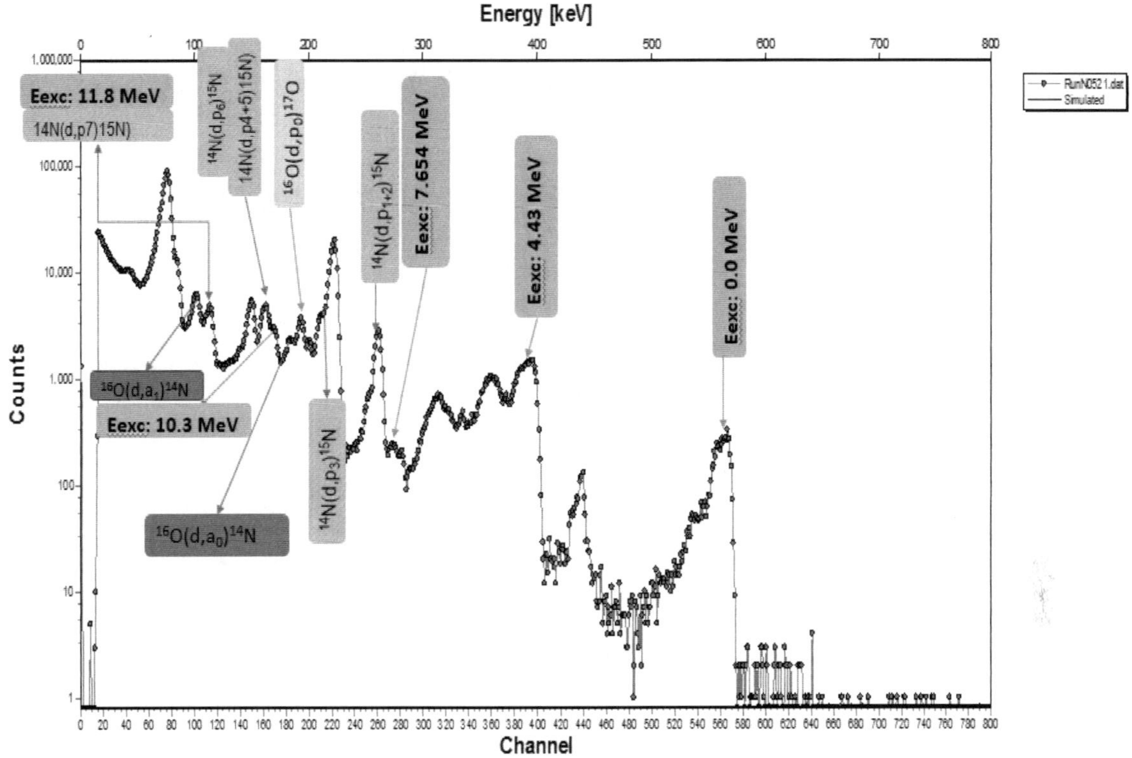

Figure 3. Spectrum obtained in the experiment with C_5H_5N target, θ_{lab}: 150°, Elab: 2.75 MeV. In the spectrum the blue boxes correspond to the reaction of interest 14N (d, a) 12C, and the excitation energies of the populated states in this spectrum are specified.

3.1 DWBA Analysis and Optical Model Potentials.

The nucleon-nucleus interaction is inherently complicated and the optical model becomes a phenomenological simple model to understand it. Potential and geometrical parameters are adjusted to fit the experimental data to extract information about the nucleus shape and structure.

The optical potential has an imaginary part that takes into account the absorption of the reaction flux from the elastic channel. This is analogous to the scattering and absorption of light by a medium of complex refractive index, which is why it is called the optical model.

Optical potentials obtained through fits of elastic scattering data are widely used to generate the distorted waves used to deduce cross-sections. Such analyses have proved to be a powerful tool in determining nuclear structures [16].

In the Distorted-Wave Born approximation, the relative motion before and after the non-elastic event is described by waves distorted by elastic scattering and its accompanying absorption. Hence this approximation is likely to be valid when the most important single event to take place when two nuclei collide is elastic scattering. Then the others reactions can be treated as perturbations, as weak transitions between elastic scattering states. [17]

Essentially DWBA calculates the cross section to a "pure" single-particle state in independent particular model, in reality, residual interactions mix these states to produce more complex wave functions, so a probability amplitude of a transfer reaction as in our case (d,α) via a particular nlj will be reduced by a factor Sjl, called the spectroscopic factor, that reveal how less "pure" than a singleparticle state is a meausered cross section.

DWBA calculation were performed using the code DWUCK4, with the optical potentials the values provided by V. Gomes Porto [18], they have the following form:

$$V(r) = Vc(r) - Vf(r) - iWg(r)$$

$$f(r) = \frac{1}{1 + exp\left(\frac{r-R}{a}\right)}$$

$$g(r) = -4a\frac{df(r)}{dr} = \frac{4exp\left[\frac{r-R}{a}\right]}{\{1 + exp\left[\frac{r-R}{a}\right]}$$

Where Vc is the potential due to a uniformly charged sphere, U and W are the depths of the real and imaginary potentials; f(r) and g(r) are Woods-Saxon and its derivative form factors respectively. Figure 4 represent the potentials used in the preliminary results.

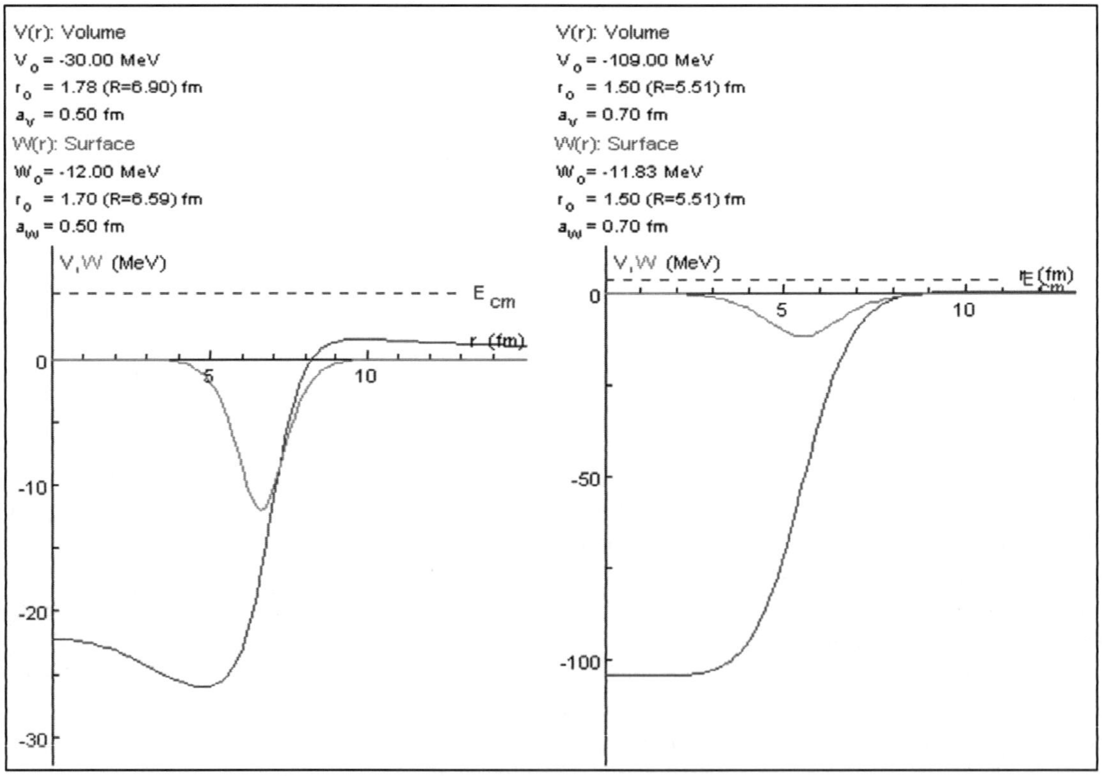

Figure 4. Optical model potentials used in the DWBA calculations performed.

When nuclear reactions occur, rearrangement of the internal structure of a nucleus will generate states of motion with more energy than the minimum the system can accommodate. These excited states are characterised by their excitation energy, spin, parity, as well as other, often only approximately, good quantum numbers. This rearrangement can easily be visualized as changes in individual nucleon motion, but can also be viewed as changes in bulk motion of the nucleus such as rotation and surface vibration.

Figures 5-8 show the experimental angular distributions obtained compared to the DWBA calculation. In every case the calculation was made twice for two possible angular momentum transfer values. The experimental values are consistent with the theoretical results.
In all cases, theoretical results were normalised to the experimental data (no spectroscopic factors).

In further analysis the spectroscopic factors will be calculated.

In the case of the Figure 5, is easy to see that definitely the transferred angular momentum value is *l*=2, in the results only the value obtained at 125° seems to be out to the DWBA curve.

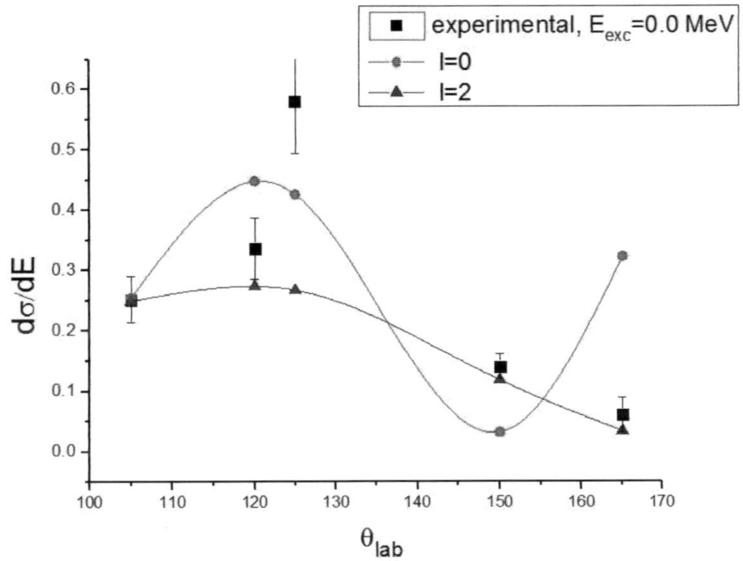

Figure 5. Angular distribution for ground state (0.0 MeV) in ^{12}C compared to the preliminary results of the DWBA analysis for l_{tr}=0 and l_{tr}=2

In Figure 6 and 7 the angular distributions for the 4.43MeV and the 7.654 MeV excited state are shown respectively, in the case of the first excited state the transferred angular momentum is *l=0*, and for the second state clearly the experimental results coincide with the *l=2* distribution.

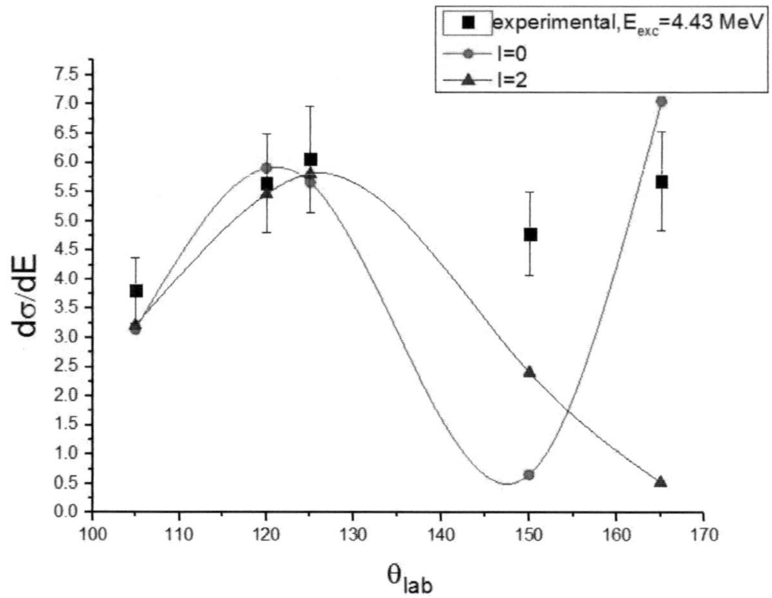

Figure 6. Angular distribution for the 4.43 MeV excited state in ^{12}C compared to the preliminary results of the DWBA analysis for $l_{tr}=0$ and $l_{tr}=2$

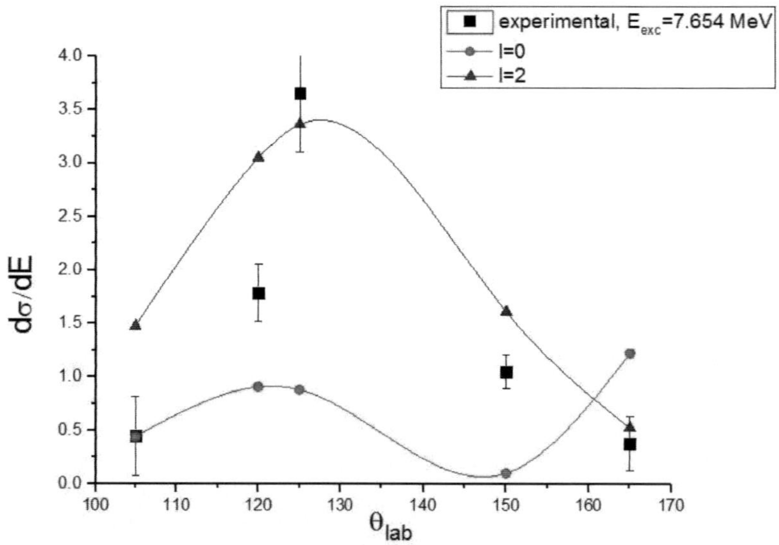

Figure7. Angular distribution for Hoyle state (7.654 MeV) compared to the preliminary results of the DWBA analysis for ltr=0 and ltr=2

In the 9.641 MeV excited state analysis, our results seem to follow the shape corresponding to an angular momentum transfer of one (*l=1*), except for the data point at 165° which has a value out of the prediction of the DWBA.

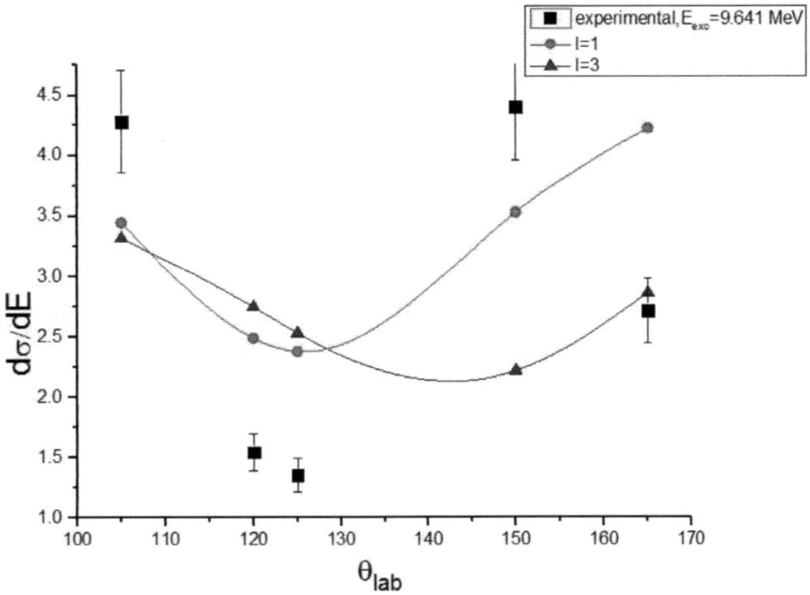

Figure 8. Angular distribution for the 9.641MeV excited state in ^{12}C compared to the preliminary results of the DWBA analysis for l_{tr}=1 and l_{tr}=3

Table 1 summarize the quantum numbers assigned after the adenine target experiment.

	0.0 MeV	4.43 MeV	7.65 MeV	9.64 MeV
ltr	2	0	2	1
J^{π}	0+	2+	0+	3-

Table 1. Values of transfered angular momentun assigned and the values of Jπ used in the DWBA analysis.

Conclusions

In this work, we presented the preliminary results of the study of some excited states in ^{12}C performed at the tandem accelerator facility at the National Institute of Nuclear Research; angular distributions were presented and discussed.

Several resonances were found in the experiment with the silicon nitride target and a medium resolution in the peaks of interest was obtained in the experiment with the adenine target.

DWBA calculations were made to compare the predicted angular distribution predictions to the data and be familiar with the process of assigning quantum numbers to the observed states.

At this stage, no attempt was made to deduce spectroscopic factors, this is left as part of our future work.

IOP Conf. Series: Journal of Physics: Conf. Series **876** (2017) 012015 doi:10.1088/1742-6596/876/1/012015

In order to obtain better data, it would be ideal to perform the next experiment with the supersonic gas jet target (SUGAR) at the 5.5 MV Van de Graaff accelerator at the Physics Institute of the UNAM [19], where a study of the same nuclear reaction was performed recently. In those experiments good statistics and resolution were obtained simultaneously, for both, the low-lying excitation region and the higher region in ^{12}C, important for understanding the nuclear structure and behaviour of the rotational band of the Hoyle state.Such experiments are planned to occur in the early summer 2017.

Acknowledgments
The authors would like to thank the EN-Tandem Accelerator operator P. Villaseñor. This work was possible thanks to the financial support of CONACYT through the numbers of the subsidies: 123655, 271802 and DGAPAUNAM: IN-115213, IG-101016.

References
[1] E. E. Salpeter, Astrophys. J. 115, 326 (1952).
[2] S.Pellegrino, Nucl Instrum Methods Phys Res B 219-220 (2004) 140-144.
[3] A.Gurbich, Nucl Instrum Methods Phys Res B 266 (2008) 1206-1208.
[4] M. Chernykh *et al*, On the Structure of the Hoyle State in 12C, PRL 98, 032501 (2007).
[5] F. Hoyle, Astrophys. J. Suppl. 1, 12 (1954).
[6] H. O.U. Fynbo *et al.*, Nature 433, 136 (2005).
[7] M. Kamimura *et al*, Nucl. Phys. A 351, 456 (1981).
[8] E. Uegaki, S. Okabe, Y. Abe and H. Tanaka, Prog. Theor. Phys. 57, 1262 (1977).
[9] M. Chernykh, H. Feldmeier, T. Neff, P. von Neumann Cosel and A. Richter, Phys. Rev. Lett. 98, 032501 (2007).
[10] A. Tohsaki, H. Horiuchi, P. Schuck, and G. Ropke, Phys. Rev. Lett. 87, 192501 (2001).
[11] D.J Marín-Lambarri *et al*, PRL 113, 012502 (2014).
[12] R. Bijker and F. Iachello, Phys. Rev. C 61, 067305 (2000).
[13] R. Bijker and F. Iachello, Ann. Phys. (Amsterdam) 298, 334 (2002).
[14] M. Mayer, "SIMNRA User's Guide", Tech. Report IPP, 9/113, Rep. IPP, p. 9/113, 1997.
[15] P. D. Kunz, manual for code DWUCK4 "Zero Range Distorted Wave Born Approximation".
[16] Hodgson, P.E. (1997). The nuclear optical model introductory overview. Nuclear Energy Agency of the OECD (NEA): OECD Publications.
[17] G.R. Sachtler, Introduction to nuclear reactions 2nd edition.1990, Oxford University Press.
[18] V. Gomes Porto, Nuclear Physics A136 (1969) 385-413.
[19] F. Favela *et al*, PRST—AB 18, 123502 (2015).

IOP Conf. Series: Journal of Physics: Conf. Series **876** (2017) 012016 doi:10.1088/1742-6596/876/1/012016

Main Nuclear Physics requirements for the robustness of r-process nucleosynthesis calculations in slow ejecta from neutron-star mergers (NSM)

J J Mendoza-Temis[1,2] **and A Frank**[1,2]

[1]Instituto de Ciencias Nucleares, Universidad Nacional Autónoma de México, 04510 México, D.F., México

[2]Centro de Ciencias de la Complejidad, Universidad Nacional Autónoma de México, 04510 México, D.F., México

E-mail: joel.mendoza@correo.nucleares.unam.mx

Abstract. Here we deal with r-process nucleosynthesis simulations for matter ejected dynamically in NSM, recently a number of such simulations (see [1, 2, 3, 4]) have display a robust pattern in their final yields. It is the main goal of this contribution to address at the main requirements from the nuclear physics point of view in order to guarantee a robust pattern in the final r-process abundances for mass numbers $A > 120$. Our results suggest that one can achieve such behaviour for slow ejecta from neutron-star mergers as long as fission cycling is involved. On the other hand, using a representative r-process calculation as a working example, we explored the main stages in the evolution of an r-process. Finally, we conclude that fine tunning of local features in the pattern of final yields (2nd, rare earth and 3rd r-process peaks) depend on an interplay between the mass surface and other nuclear structure properties involved.

1. Introduction

There is no doubt that one of the most interesting open questions is that of how nature builds its elements. Nuclear structure properties, in particular nuclear masses, particle capture rates, β-decay and fission are the most important ingredients to elucidate this question. The so-called rapid neutron-capture process (r-process), is known to be at the origin of approximately half of the $A > 60$ stable nuclei observed in nature. It is commonly accepted that the r-process occurs as a sequence of neutron captures and β-decays in environments with extreme neutron densities, under such physical conditions matter reaches a region farther away from stability, which makes the calculation of the r-process harder that it seems, because so far only a few of the required nuclear structure properties are experimentally known, consecuently almost the entire set have to rely on theoretical grounds. And, on top of that, the astrophysical site of the r-process remains unindentified. For many years the neutrino-driven wind from the surface of a freshly born neutron star in a corecollapse supernova has been the favored site [5]. However, recent supernova simulations [6] indicate that the conditions of the matter ejected in the wind (entropy, expansion velocity, proton-to-neutron ratio) are not suited to support an r-process which produces the elements in the third peak (around mass number A$\sim$195) and beyond. We have performed r-process simulations based on recent numerical simulations of NSM, which indicate that the matter ejected during the dynamical phase is very neutron rich with extremely large neutron-to-seed ratios ($R_{n/s} > 400$) [1, 2, 3]; i.e. there are many neutrons which can be captured by seed nuclei transporting matter to very heavy nuclei in the region of the nuclear chart where decay by fission is possible. Our work is organized as follows. In the next section we give a brief description the main nuclear physics input required to performed r-process calculations. Then in section 3, we introduce a working example to understand the relavant stages in evolution of an r-process from the nuclear physics point of view. The Results of our simulations are presented and discussed in section 4. Finally, we conclude in section 5.

2. Nuclear physics input

Nuclei are particularly complex systems, ranging from a single proton to more than three hundred nucleons; too large for a detailed microscopic treatment but too small for statistical methods. Interest in them is not restricted to nuclear physicists because the atomic nuclei that constitute our world are manufactured inside stars, defining their evolution and fate. In this contribution we are mainly interested in performing r-process calculations. The main nuclear physics properties required to achive our goal are: nuclear masses, reaction rates, such as n-capture (n, γ), phodissociation (γ, n), β-decay, and rates for neutron induced and β-delayed fission, as well as the corresponding fission fragment distributions. In what follows, a brief description of each of them will be provided.

2.1. Nuclear masses

Arguably, the most basic property of a nucleus is its mass. Understanding nuclear masses provides a test of our basic knowledge of nuclear structure and is an essential ingredient of the fundamental astrophysical problem of nucleosynthesis, which often takes place in far-from-stability conditions, on ultrashort time scales [7]. Though great progress has been made on measurement the mass of short-lived nuclei that are far from the region of stable, naturally occurring isotopes, theory is needed to predict their properties and guide experiments that search, for example, for regions of increased stability [8]. Even when most of the r-process nuclei are not yet reachable by up-today facilities, experimental mesurements are still needed, on the one hand for validating theoretical mass models, and on the other as inputs for r-process models. The efforts to calculate nuclear masses have been hampered by the absence of a true effective theory of the nuclear interaction and by the difficulties inherent to quantum manybody calculations. Instead, simplified approaches to model the atomic nucleus have been devised. The predictions made by different models often turn out to agree in regions where masses are available, but diverge profoundly away from known regions. Reliable theoretical models and methodologies that can predict the mass and other properties of these exotic nuclei are still missing [9]. In summary, Nuclear masses are particularly important as they, via the neutron separation energies, define the r-process path in the nuclear chart and secondly they are crucial ingredients in the statistical model calculation of neutron capture cross sections. The impact of a number of theoretical nuclear mass models (FRDM [10], HFB21 [11], WS3 [12], DZ31 [13, 14] and range of possible Skyrme functionals within the DFT framework) for the calculation of the neutron capture rates that enter in the r-process simulations has been widely explored in previous works (see [4, 15] for more details). Another approach is through Monte Carlo simulations, in which nuclear masses are randomly varied within experimental or estimated theoretical uncertainties, such studies suggest that in order for details of the abundance patterns to stand out over nuclear uncertainties, a mass precision of 100 keV. [16]. Unfornately, the overall RMS deviation between theoretical and experimental masses for the range A=16-270 is 500 keV or greater. As the descripton of the so called "deformed nuclei" is one of the main sources of discrepancies among all theoretical mass models, and as their predictions are relevant for r-process calculations, in what follows we will briefly discuss this issue.

2.1.1. The region of transition from spherical to deformed nuclei. Regardless of the theoretical approach used for the calculation of the nuclear masses and important feature that all the models shared is the deformation. To display the predicted "deformed nuclei region" for a number of selected mass models, we are going to use the two neutron separation energies ($S_{2n}/2$), because it is well known, that they mostly display an smooth behaviour except in regions where shell closures or the onset of deformation is present. In Fig. 1, we display $S_{2n}/2$ as a function of the mass number A for isotopic chains ranging from $30 < Z < 92$. The experimental values of $S_{2n}/2$ are always included when available, and displayed by red lines. In the absence of experimental information, the theoretical values are displayed by black lines. The predicted deformed regions are added and identified by green lines on top of the theoretical value of $S_{2n}/2$. Finally every 5 isotopic chains bluish lines are included only to guide the eye. To stress the need of theoretical predictions and as a reminder, let us consider the waiting point approximation [17], assuming typical r-process conditions, i.e., $T_9 = 1$ GK and $n_n = 10^{24}\ cm^{-3}$, we obtain an r-process path of $S_n^0 = S_{2n}/2 \approx 2$ MeV, meaning that most of relevant nuclei for r-process calculations lie not only in the extrapolated region but also in the "deformed region" (see purple box region in Fig. 1). Concerning to the predicted $S_{2n}/2$ surfaces (see black and green lines in Fig. 1), the first thing to glimpse the eye is the extremely irregular behavior of the predicted surfaces by HFB21 and FRDM in comparison to those of WS3 and DZ31. In addition, in between the shell closures (which in this representation appear as vertical lines) there are two regions, namely for $90 < A < 110$ and for $130 < A < 170$, where most of the selected models predict an onset of deformation (see green lines in Fig. 1). From the nuclear physics point of view, this piece of information is of vital relevance to understand the evolution of an r-process.

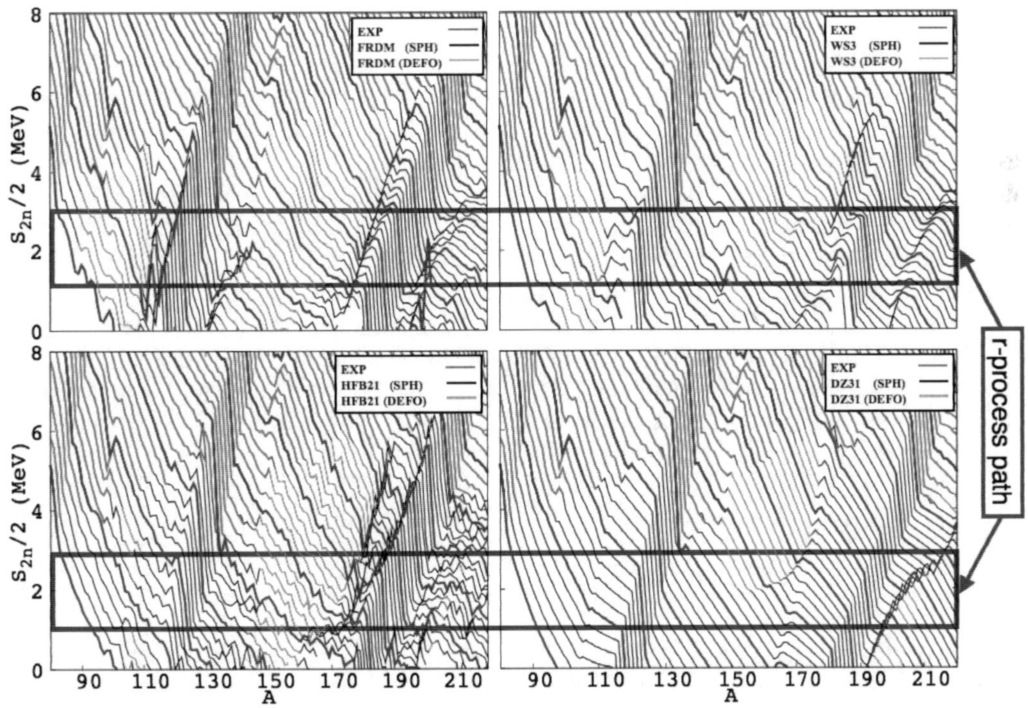

Figure 1: (Color online) $S_{2n}/2$ for a number of isotopic chains ($30 < Z < 92$) based on a number of mass models (see black lines). Experimental data is always shown with red lines. Well deformed nuclei are displayed with green lines. Blue lines are added every five isotopic chains to guide the eye. Relevant nuclei on the r-process path are denoted by a purple box.

2.2. Reaction rates
A key ingredient in any astrophysical calculations is the knowledge of the thermonuclear reaction rates via their cross sections; as they measure the probability of reactions between a target nucleus and incoming projectiles under astrophysical conditions. The commonly used tool to generate theoretical reaction rates for r-process rely on the Hauser-Feshbach theory [18], where all energetically possible reaction channels are in statistical competition with a probability evaluated by dividing over the sum of transmission coefficients for all channels as in Eq. 1, where T is the transmission coefficient, J^{π} the spin and parity of each populated level, and E is the energy at which the reaction is evaluated. The subscript c denotes the formation of a compound nucleus while the index i runs over all possible exit channels for the decay of the compound system.

$$\sigma(E) = \Sigma_{J^{\pi}} \frac{T_c(J^{\pi})T_{\gamma}(J^{\pi})}{\Sigma_i T_i(J^{\pi})} \tag{1}$$

This picture, is suitable for the description of reactions involving the decay of nuclei excited at a sufficiently high energy to contain a large number of levels per MeV. When this condition is true an energy-averaged reaction cross section in this region of highly overlapping resonances is meaningful. Assuming that this condition is met, the main sources of physics-related uncertainty in the calculated reaction rates come from the modeling of the formation and decay of the excited compound nucleus through the respective transmission coefficients. These are calculated using nuclear level densities to describe the excitation of each compound nucleus, optical potentials to describe the emission or absorption of particles, and γ-ray strength functions to describe the emission of γ-rays [19].

2.3. β-decays
Besides masses and reaction rates, the β-decay half-lives are an important physical ingredient to perform r-process nucleosynthesis simulations. They must be known all over the nuclear chart, but particularly in the region between the valley of stability and the r-process equilibrium path. While masses determine the route of r-process under given astrophysical conditions, the β-decay half-lives of waiting-point nuclei dictate how fast the process can proceed towards heavy nuclei and how much material is accumulated in a given isotopic chain. In other words, the effective r-process timescale can be affected by modifications of individual β-decay half-lives that so far are based on relatively uncertain theoretical approaches [20, 21]. As an example, of the relevance of β-decays half-lives, the competition between β-decay and n-capture (n, γ) in the region A$\sim$160 during freeze-out stage can have significant consequences for the formation of a small but pronounced rare-earth peak [22, 23]. In addition, during freeze-out and because of neutron capture rates are not fast anymore, β-delayed neutron emission takes place changing the isobaric decay chain towards stability.

2.4. Fission
Fission rates and fragment distribution of neutron-rich heavy nuclei are required to understand r-process nucleosynthesis, specially at late-time r-process dynamics through the so-called recycling mechanism [24, 25]. Under NSM conditions fission cycling is unavoidable due to the large neutron-to-seed ratios ($R_{n/s}$) reached in their ejecta, in addition fission is also an important source of free neutrons, under r-process conditions the total number of neutrons produced can be divided into two components: 1) a prompt component that consists of the neutrons evaporated mainly by the highly excited fragments, and 2) a delayed component that occurs during the decay of the fragments to the instantaneous r-process path. Unfortunately, a comprehensive, microscopic explanation of nuclear fission still eludes us due to the complexity of the process. This fundamental nuclear decay is an example of a quantal large-amplitude collective motion [26].

3. Numerical Simulations. Description of a working example.

Previously in [4], we have performed a number of simulations to explore the impact of nuclear masses in r-process nucleosynthesis, covering the full range of ejecta coming from a 3D simulation of NSM [27]. As the current work is intended to understand the nuclear physics occurring at relevant stages of the evolution of an r-process, we decided to use a working example, in which a representative trajectory for slow ejecta from NSM has been taken, such trajectory characterizes by being highly neutron-rich ($Y_e \sim 0.05$), and having a neutron-to-seed ration $R_{n/s} \sim 650$, favouring an initial composition from Nuclear Statistical Equilibium (NSE) centering at $A \sim 120$. Concerning the Nuclear physics input, our calculations consider a nuclear network already available [4], where as nuclear reactions we considered charge particle reactions, neutron captures and its inverse process, photo-dissociation, and β and α-decay and fission. In [4, 28], we have derived the neutron capture rates consistently for a number of mass model within the statistical model using the code MOD-Smoker (see [30] which includes changes and updates from the code NONSMOKER due to Rauscher and Thielemann [31]), along this working example, we decided to choose the theoretical binding energies from DZ31, because the physics behind this model is the more familiar to the authors (see [14]), and most important, we prefer to deal with DZ31 mass predictions that display a smoother mass surface far from stability (see Fig. 1) than with a noisy mass surface (as predicted by most models) which could be attributed to numerical noise rather than physics. The photodissociation rates were obtained from the neutron capture rates by detailed balance. For nuclei, for which the half lives are not known experimentally, we have adopted the β-decay (and β-delayed neutron emission) rates from the compilation of Möller *et al.* [20], which was derived from QRPA calculations on top of the FRDM mass model. We used the parametrization of Ref. [32] of the Viola-Seaborg formula to estimate the α-decay rates, which become relevant for heavy nuclei beyond lead. Finally, for nuclei with $Z > 83$, where a competition between (n, γ) and neutron induced fission can take place, we used neutron-induced reaction rates taken from [33] that are based on the FRDM mass model [10] and the Thomas-Fermi fission barriers of Myers and Swiatecki [34]. Rates for β-delayed and spontaneous fission were adopted from [35]. Our fission yields were taken from the calculations of Ref. [36] which were derived using the code ABLA. This approach also gives a consistent estimate for the number of neutrons set free during the fission process.

4. Results at relevant stages of an r-process

Figs. 2 to 4, shown r-process nucleosynthesis results for our prevously described working example at 3 different phases of the evolution :

(a) at freeze-out, which we define as the moment where $R_{n/s} = 1$,

(b) at the moment when β-decays becomes the dominant process ($\tau_{(n,\gamma)} >> \tau_\beta$) as there are not enough neutrons ($R_{n/s} \sim 10^{-10}$), and

(c) at the final abundance, calculated at a time of 1 Gyr.

To better interpet our findings, we have decided to plot r-process abundaces (red lines for DZ31, and blue dots for solar) as a function of atomic number A (see upper panel on l.h.s. in Figs. 2 to 4), and a color coded version of the same on the N-Z plane (see r.h.s. in Figs. 2 to 4, where stable isotopes are addead as black empty boxes). To understand the relevant nuclear physics, we added the instantaneous r-process path, on top of the $S_{2n}/2$ (previouly described in Fig. 1), where black dots represent the most abundant isotope for a given isotopic chain Z (see lower panel on l.h.s. in Figs. 2 to 4). In addition to discuss the relevant regions of the landscape, where the evolution of an r-process is taking place we added circles with numbers inside.

4.1. freeze-out

As initially matter is highly neutron rich, the dominant process is that of n-capture. When the *r*-process reaches the freeze-out, the abundances show a strong odd-even staggering (see upper panel in Fig. 2), matter is mostly halted at neutron shell closures (see lower panel in Fig. 2), namely at N=126 (A∼190) and N=184 (A∼280), which translates in troughs in the abundance pattern right after those major shell closures. The amount of material accumulated in the fissioning region, A>240 is much larger than the one present in the region below the 3rd *r*-process peak, this occurs because the flux of matter is determine by β-decay half-lives, and as τ_β(N=184)>τ_β(N=126) matter stays longer in the transuranium region (A>240). There are 3 regions of interest:

1. **Region between 2nd and 3rd *r*-process peak (130<A<180).** As it is shown by the *r*-process path (black dots in lower panel in Fig. 2), the odd-even staggering in abundances are due to local effects induced by a sudden change in the intrinsic deformation starting at N=90 (A∼140) (green lines in lower panel in Fig. 2) and the onset of a sub-shell at N=112 (A∼160) which produces another trough in the abundances. The latter effects depends on the mass model.
2. **Region after 3rd *r*-process peak (200<A<240).** Local features in the abundances are explained in the same way as in item 1. In this region, deformation sets in at N=130 (A∼200), and the sub-shell is located at N∼150 (A∼230).
3. **Transuranium region (A>240).** There is a sub-shell at N∼170 (A∼260) which produces a notorius trough in abundances. Nuclei located at N=184 (A∼280) have lower barriers and consecuently n-induced fission dominates over n-capture, and the *r*-process cycles to medium mass nuclei rather than producing heavier nuclei.

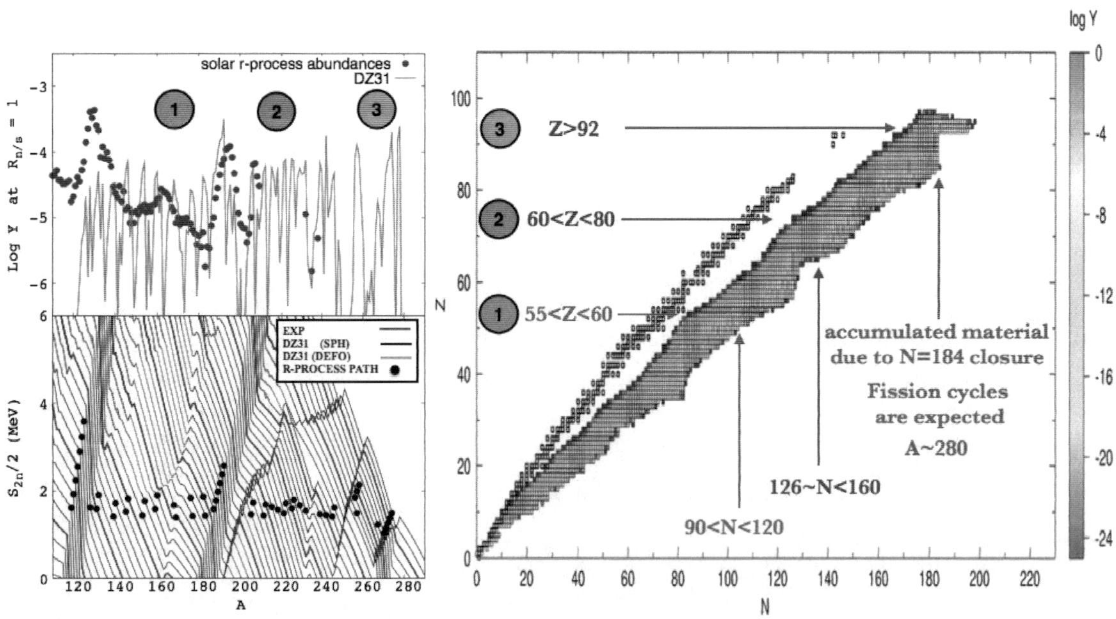

Figure 2: (Color online) Red lines display r-process abundances as a function of mass number at freeze-out stage, for a representative slow ejecta from NSM and masses taken from DZ31, solar abundances as blue dots for comparison (see upper panel l.h.s.). $S_{2n}/2$ surfaces, as described in fig. 1, and on top of them black dots representing the r-process path (see lower panel l.h.s.). Color coded Abundances on N-Z landscape and stable isotopes as black empty boxes (see r.h.s.).

4.2. beta-decay phase

Once neutron captures are slower than β-decays, matter decays to stability. In particular, at this stage, the remarkable features of the r-process abundance pattern, namely the 2nd peak, the rare earth peak, and the 3rd peak appear almost at the right height and place (see upper panel on l.h.s in Fig. 3). There are 5 regions to be consider:

1. **2nd r-process peak (A$\sim$130).** The fission fragment ditribution for nuclei at A$\sim$280 predicted by ABLA code [36] (see r.h.s in Fig. 3) shows that the origin of this peak is mainly due to fission cycling, as one of its fragments, precisely lie at A$\sim$130. The position of the peak is slighly shifted to the right due to late-time n-captures.
2. **Region around (A$\sim$140).** This kink in the abundances is produced both, by matter halted at N=82 and by fission, as the 2nd fission fragment for nuclei at A$\sim$280 predicted by ABLA code [36] mostly lie at A$\sim$140.
3. **rare earth peak (A$\sim$160).** This peak is mainly produced by the competition between β-decay and n-capture (n, γ) in the region A$\sim$160 soon after the freeze-out [22, 23]. Our results suggest, that at this stage, the instantaneous r-process path at A$\sim$160 lies in the region of transition from spherical to deformed nuclei (green lines in lower panel on l.h.s. in Fig. 3), which makes this peak very sensitive to the mass model being used.
4. **3rd r-process peak (A$\sim$195).** The origin of this peak is matter halted at N=126 expose to late-time n-captures. It turned-out that this peak is wider than expected, because there is a sub-shell effect at N=112 (A$\sim$180) which is mainly accumulating material. Again, our findings suggest that this peak is very sensitive to the mass model being used.
5. **Region above Lead A>208.** The significant amount of matter above lead still existing (see upper panel on l.h.s. in Fig. 3 and on r.h.s. in Fig. 4), will mostly decays via α-decay to finally form the lead peak. To estimate the α-decay rates, we have made used of [32].

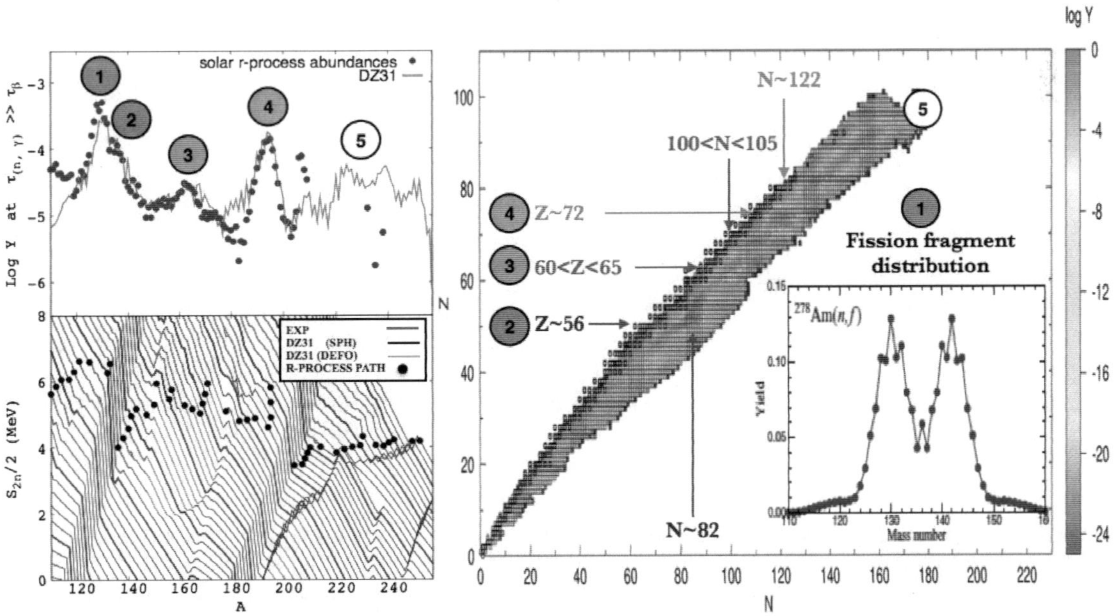

Figure 3: (Color online) r-process abundances when β-decay becomes the dominant process, same description as Fig. 2. Fission fragment distributions as function of mass number for neutron induced fission on ^{278}Am as predicted by the ABLA code are added on r.h.s.

4.3. final abundances at 1GY

We noticed that the final abundances at 1 Gyr are smoother due to late-time n-captures coming from fission and subsequent β-decay and photodissociation. When comparing with solar abundances, we noticed that overall the main features of r-process abundances pattern at 1 GY are well reproduced (see upper panel on l.h.s in Fig 4). The slighly differences at the 2nd, at the rare earth peak and at the 3rd r-process peak has been explained in subsection 4.2. The color coded version of the abundances on the N-Z plane shows nuclei responsible for the above mentioned r-process peaks (see r.h.s. in Fig. 4). Finally, there is an over-estimation in both the Lead Peak (A$\sim$208) and in the abundances associated with the long-lived Thorium and Uranium isotopes (A>230).

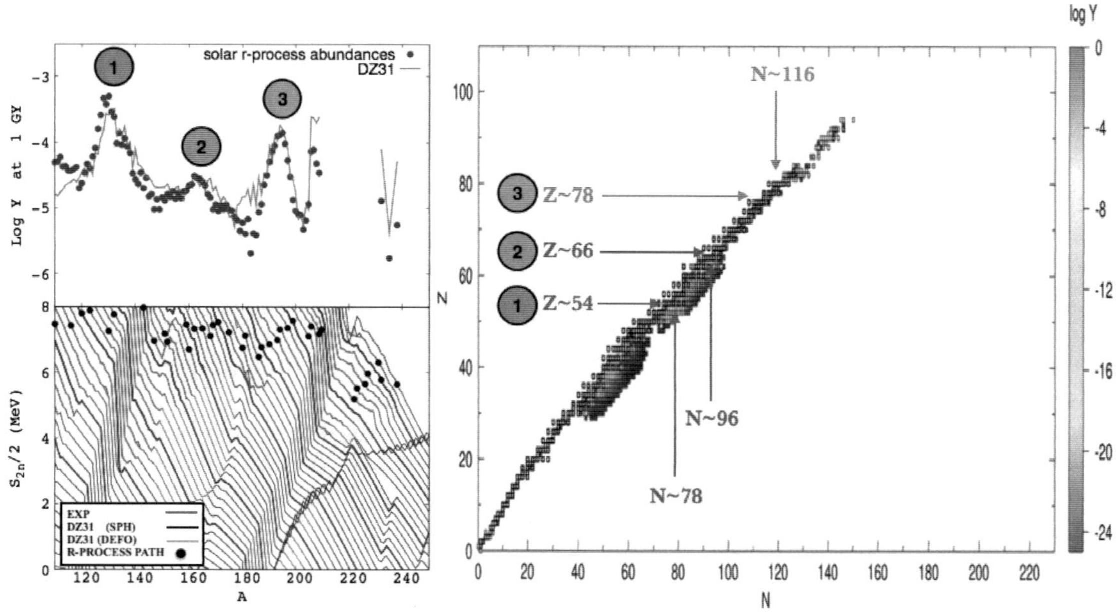

Figure 4: (Color online) r-process abundances at 1 GY, same description as Fig. 2

4.4. On the robustness of r-process abundances

Up to now, we have only dealt with our working example to gain insight on relevant nuclear physics at different stages of the evolution of an r-process, however a remarkable result we have found in [4] is worth to be mentioned: "...When considered only the so-called "slow ejecta", the most striking feature of our calculations is the fact that the final abundances for mass numbers A>120 are virtually identical, for a given mass model." (see Fig. 5). In what follows a few words on the impact of nuclear masses in the final r-process abundances would be stated.

As we have noticed the 2nd r-process peak is due to fission fragments coming from A$\sim$280 hence its origin is independet of the mass model being used; soon afterwards neutron captures on the fission yields are responsible for a flow of matter from the 2nd r-process peak to heavier nuclei. This mechanism operates in all used mass models except in FRDM due to the fact that material is halted at N$\sim$90, this feature translates in a notorious over-estimation of the 2nd r-process peak, which is also shifted to the right (see upper leftmost panel in Fig 5).

To understad the behavior around the 3rd r-process peak, table 1 shows the one-neutron separation energies (S_{1n}) predicted by a number of mass models for the nuclei ^{199}Yb, ^{198}Tm and ^{197}Er (all with N=129, just above the magic number N=126). The FRDM and HFB21 mass

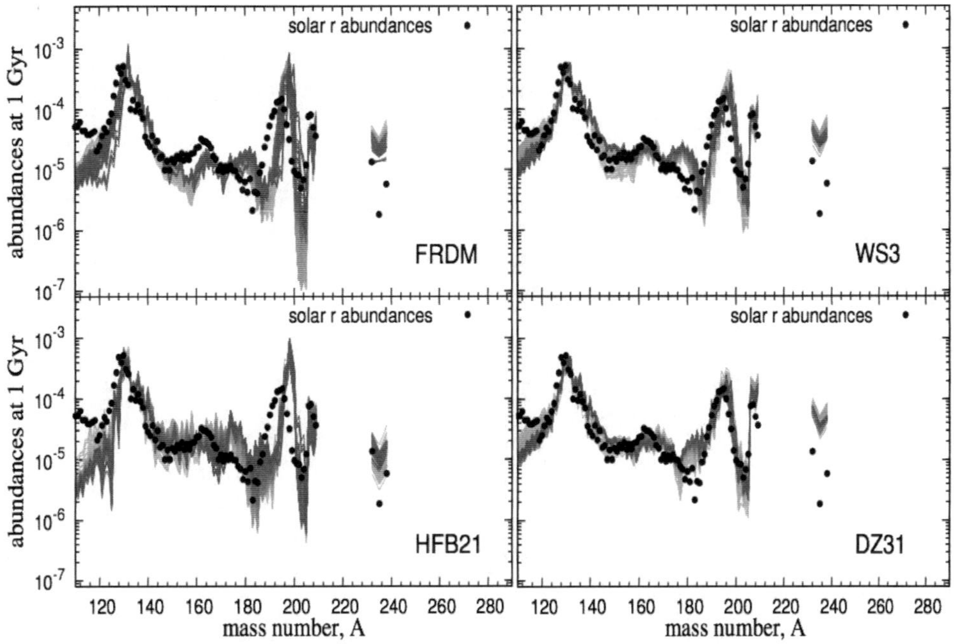

Figure 5: (Color online) Final r-process abundances at a time of 1 Gyr for the different mass models and all the slow ejecta trajectories.

models predict noticeably smaller neutron separation energies than the DZ31 or the WS3 models in this mass range. Thus these nuclei act as (additional) obstacles in r-process simulations using the FRDM and HFB21 mass models, even if the mass flow has overcome the $N = 126$ waiting points. As a result, the third peak in the abundance distribution is shifted for these two mass models to higher mass numbers as can be seen in Fig. 5 caused mainly by late-time neutron captures. Due to the larger neutron separation energies, the $N = 129$ nuclei do not act as obstacles in simulations adopting the Duflo-Zuker or WS3 mass models. Relatedly the third peak develops at $A \sim 195$, associated with the $N = 126$ waiting points.

Table 1: S_{1n} (in MeV) for a number of selected mass models, for the region N=129.

	^{199}Yb	^{198}Tm	^{197}Er
FRDM	0.52	0.62	0.26
HFB21	0.85	0.73	0.56
WS3	1.46	1.53	0.86
DZ31	1.39	1.48	0.81

5. Conclusions

We conclude that dynamical ejecta of NS mergers show a robust r-process pattern, as already concluded in refs. [1, 2, 3, 4], provided that the ejecta remains neutron rich, see refs [38, 39, 40]. The main requirement to achieve a robust r-process pattern is that the amount of material accumulated at freeze-out in the fissioning region, A$\gtrsim$240 is much larger than the one present in the region below the 3rd r-process peak. For the slow ejecta, this is guaranteed by the fact that the beta-decay half-lives grow with increasing mass number and by the presence of a neutron shell closure around $N = 184$. Both effects are responsible of producing a peak in the freeze-out

r-process abundances around A$\sim$280 (see upper panel on l.h.s. in Fig. 2). The material in this peak decays by fission contributing to the abundances around the 2nd r-process peak and producing a final robust r-process pattern. The rare earth peak and the 3rd peak abundances are noticeably more sensitive to nuclear masses that influence both the n-capture rates and the β-decay rates. It is not surprising that abundances of these peaks show a larger variation.

Acknowledgments
J.J. Mendoza-Temis is a DGAPA-UNAM fellow and thanks to DGAPA-PAPIIT project number IV100116 for its support.

References
[1] S Goriely, A Bauswein, and H-T Janka, Astrophys. J. **738**, L32 (2011).
[2] O Korobkin, S Rosswog, A Arcones, and C Winteler, Mon. Not. R. Astron. Soc. **426**, 1940 (2012).
[3] A Bauswein, S Goriely, and H-T Janka, Astrophys. J. **773**, 78 (2013).
[4] J J Mendoza-Temis, M R Wu, K Langanke, G Martínez-Pinedo, A Bauswein, and H-T Janka, Phys. Rev. C **92**, 055805 (2015).
[5] S. E. Woosley, J. R. Wilson, G. J. Mathews, R. D. Hoffman, and B. S. Meyer, Astrophys. J. **433**, 229 (1994).
[6] H.-T. Janka, Ann. Rev. of Nucl. Part. Sci. **62**, 407 (2012).
[7] C. E. Rolfs and W. S. Rodney, *Cauldrons in the Cosmos* (University of Chicago Press, Chicago, 1988).
[8] Y. Oganessian, Nature (London) **413**, 122 (2001).
[9] D. Lunney, J. M. Pearson, and C. Thibault, Rev. Mod. Phys. **75**, 1021 (2003).
[10] P Möller, J R Nix, W D Myers, and W J Swiatecki, At. Data Nucl. Data Tables **59**, 185 (1995).
[11] S Goriely, N Chamel, and J M Pearson, Phys. Rev. C **82**, 035804 (2010).
[12] M Liu, N Wang, Y Deng, and X Wu, Phys. Rev. C **84**, 014333 (2011).
[13] J Duflo and A P Zuker, Phys. Rev. C **52**, R23 (1995).
[14] J Mendoza-Temis, J G Hirsch, A P Zuker, Nuclear Physics A **843** (1), 14-36 (2010)
[15] D. Martin, A. Arcones, W. Nazarewicz, and E. Olsen, Phys. Rev. Lett. **116**, 121101 (2016).
[16] M. Mumpower, R. Surman, G.C. McLaughlin, and A. Aprahamian, Prog. Part. Nucl. Phys **86**, 86 (2016).
[17] A G W Cameron, J J Cowan, J W Truran, *Astrophysics and Space Science*, **91** 2 (1983).
[18] W. Hauser, and H. Feshbach, Phys. Rev., **87** 2 (1952).
[19] J. Pereira, *Private communication* (2016).
[20] P Mller, B Pfeiffer, and K-L Kratz, Phys. Rev. C **67**, 055802 (2003).
[21] T. Marketin, L. Huther, and G. Martnez-Pinedo, Phys. Rev. C **93**, 025805 (2016)
[22] R. Surman, J. Engel, J. R. Bennett, and B. S. Meyer, Phys. Rev. Lett. **79**, 1809 (1997)
[23] M. R. Mumpower, G. C. McLaughlin, and R. Surman, Phys. Rev. C **85**, 045801 (2012)
[24] G. Bell, Phys. Rev. **158** (1957)
[25] S. Goriely, J.-L. Sida, J.-F. Lemaître, S. Panebianco, N. Dubray, S. Hilaire, A. Bauswein, and H.-T. Janka, Phys. Rev. Lett. **111** 24 (2013)
[26] W. Nazarewicz, *Private communication* (2016).
[27] A Bauswein, H-T Janka, and R. Oechslin, Phys. Rev. D **82**, 08404 (2010).
[28] J. Mendoza-Temis, Ph.D. thesis, TU Darmstadt, (2014).
[29] J. M. Lattimer, Annu. Rev. Nucl. Part. Sci. **62**, 485 (2012).
[30] H. P. Loens, Diploma thesis. TU Darmstadt, (2007).
[31] T. Rauscher and F.-K. Thielemann F.-K., At. Data Nucl. Data Tables **75**, (2000).
[32] T Dong and Z Ren, Eur. Phys. J. A **26**, 69 (2005).
[33] I V Panov, I Y Korneev, T Rauscher, G Martínez-Pinedo, A Kelić-Heil, N T Zinner, and F-K Thielemann, Astron. Astrophys. **513**, A61 (2010).
[34] W D Myers and W J Swiatecki, Phys. Rev. C **60**, 014606 (1999).
[35] I Petermann, K Langanke, G Martínez-Pinedo, I V Panov, P-G Reinhard, and F-K Thielemann, Eur. Phys. J. A **48**, 122 (2012).
[36] N T Zinner, Ph.D. thesis, University of Aarhus, 2007 (unpublised).
[37] A Arcones and G Martínez-Pinedo, Phys. Rev. C **83**, 045809 (2011).
[38] S Wanajo, Y Sekiguchi, N Nishimura, K Kiuchi, K Kyutoku, and M Shibata, Astrophys. J. **789**, L39 (2014).
[39] Y Sekiguchi, K Kiuchi, K Kyutoku, and M Shibata, Phys. Rev. D **91**, 064059 (2015).
[40] S Goriely, A Bauswein, O Just, E Pllumbi, and H-T Janka, Mon. Not. R. Astron. Soc. **452**, 3894 (2015).
[41] S Goriely and G Martínez-Pinedo, Nucl. Phys. A **944** (2015).

IOP Conf. Series: Journal of Physics: Conf. Series **876** (2017) 012017 doi:10.1088/1742-6596/876/1/012017

Towards the measurement of the $^{13}C(d, p)^{14}C$ cross section using AMS

S. Murillo-Morales [1], L. Barrón-Palos [2], E. Chávez[2], V. Araujo-Escalona[2]

[1] Facultad de Ciencias, Universidad Nacional Autónoma de México. 04510, Ciudad de México.
[2] Instituto de Física, Universidad Nacional Autónoma de México. 04510, Ciudad de México.

E-mail: silviamur@ciencias.unam.mx

Abstract. A plan to study the total cross section for the $^{13}C(d, p)^{14}C$ nuclear reaction has been developed for energies in the center-of-mass frame between 133 and 400 keV. The proposed experiment will use a deuterium beam (1 - 3 MeV of energy) from the Instituto de Física-UNAM 5.5 MV Van de Graaff accelerator and the produced ^{14}C will be afterwards measured by AMS technique in the LEMA-UNAM (HVEE 1 MV Tandetron). One of the main goals is to study the performance of the LEMA-UNAM facility in the cross section measurement in comparison with other data reported in the literature, measured by other techniques. In this work we present the current status of these studies. The relevance of the $^{13}C(d, p)^{14}C$ reaction in the study of compound nucleus formation as well as in some astrophysics scenarios, and the importance of the development of the AMS technique to measure cross sections of nuclear reactions of astrophysical interest in Mexico are also discussed.

1. Introduction

Knowing the cross section of nuclear reactions is important in nuclear physics because it helps to have a better understanding of nuclear structure. In nuclear astrophysics, it provides information about the processes that take place in the cosmos. Unfortunately, the cross section of many nuclear reactions at the energies of interest remain poorly known due to the challenges implicated in the measurement of very low cross sections or in the measurement of cross sections of nuclear reactions that involve radioactive nuclei. Therefore, the development of efficient techniques to measure reaction cross sections is necessary. The use of the AMS technique to determine the presence of an isotope of interest, produced by a nuclear reaction in a previous irradiation of the sample is gaining more scientific interest as a tool to study nuclear reactions with small cross sections that are difficult to measure using other techniques, besides it offers a complementary tool for studying long-lived radionuclides [1]. Several studies have been published on the use of the AMS technique to study nuclear reactions [1] [2] [3] [4].

This paper presents the preparation that is currently taking place towards the first measurement of the cross section of the $^{13}C(d, p)^{14}C$ nuclear reaction at low energies (133 - 400 keV in the center-of-mass system) using the AMS technique.

What we know about the cross section of the $^{13}C(d, p)^{14}C$ nuclear reaction is largely based on very limited experimental data . Hence, one of our goals is to widen experimental knowledge of this reaction. Previous work on the $^{13}C(d, p)^{14}C$ reaction has focused on the angular momentum distribution and yield [5][6], and the excitation function measurement [7].

This reaction involves deuterium, which is an important ingredient in the primordial nucleosynthesis that took place within the first 3 minutes after the Big Bang [8]. Although the $^{13}C(d,p)^{14}C$ reaction is not considered among the main reactions that take place in the standard CNO nucleosynthesis, extended models where subleading reaction channels are considered seem to find larger concentrations of ^{13}C and ^{14}C due to a more efficient carbon burning process that involves (d,p) reactions on ^{12}C and ^{13}C [9] .

Furthermore, this nuclear reaction is of interest since it allows to study the formation of compound nucleus. The first studies on compound nucleus formation in the $^{13}C(d,p)^{14}C$ nuclear reaction from 1 to 3 MeV in the laboratory frame of reference was carried out in 1956 by Marion and Weber [6].

Figure 1 shows the cross section for the $^{13}C(d,p)^{14}C$ nuclear reaction, which is of the order of ten mb between 1 and 3 MeV.

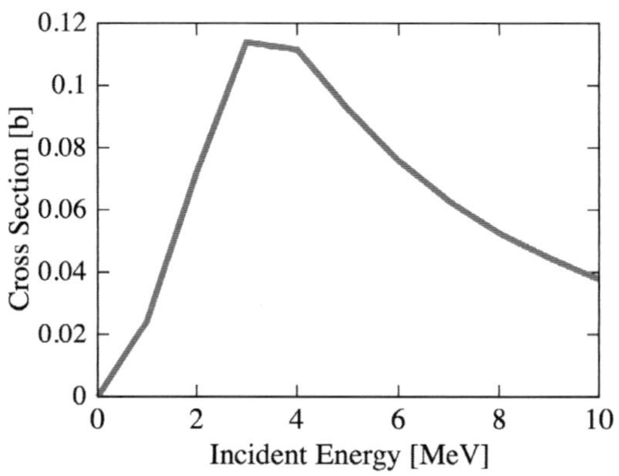

Figure 1. Total cross section for the $^{13}C(d,p)^{14}C$ nuclear reaction at energies between 0 and 10 MeV in the laboratory frame of reference. Data retrieved from ENDF (TENDL - 2014) [10].

2. Experiment

Our purpose is to irradiate graphite (98.9% ^{12}C, 1.1% ^{13}C) samples with a deuterium beam with energies between 1 and 3 MeV. In order to avoid processing the graphite after irradiation for AMS analysis, the aluminium cylindrical cathodes containing graphite, used in the negative-ion source of the AMS accelerator, are directly irradiated at the Van de Graaff accelerator. Approximately 4.5 mg of graphite are used per cathode. The surface corresponding to graphite is $\sim 1.33\text{mm}^2$. The cathode geometry is shown in Figure 2. Using two CANBERRA's Passivated Implanted Planar Silicon detectors (PIPS), Rutherford Backscattering Spectrometry (RBS) will be performed to determine how much of the sample's surface has actually been irradiated.

Despite the low content of ^{13}C in graphite, and the relatively low deuterium-beam currents that can be obtained at the IFUNAM 5.5 MV Van de Graaff ($\sim 3.12 \cdot 10^{10}$ particles/s·cm^2), it is possible to achieve relative concentrations of ^{14}C ($^{14}C/^{12}C$) that are of the order of 10^{-12} in relatively short times (a few hours). These concentrations are within the sensitivity range of LEMA, where relative concentrations as low as 10^{-15} for $^{14}C/^{12}C$ can be measured.

A plot of the calculated time variation of the relative concentration of $^{14}C/^{12}C$ is shown in Figure 3. As it can be seen, relative concentrations of 10^{-12} can be achieved after a few hours of irradiation with a 500 nA beam impinging the $1.33\,\text{mm}^2$ cross sectional area of graphite on the aluminum cathode surface.

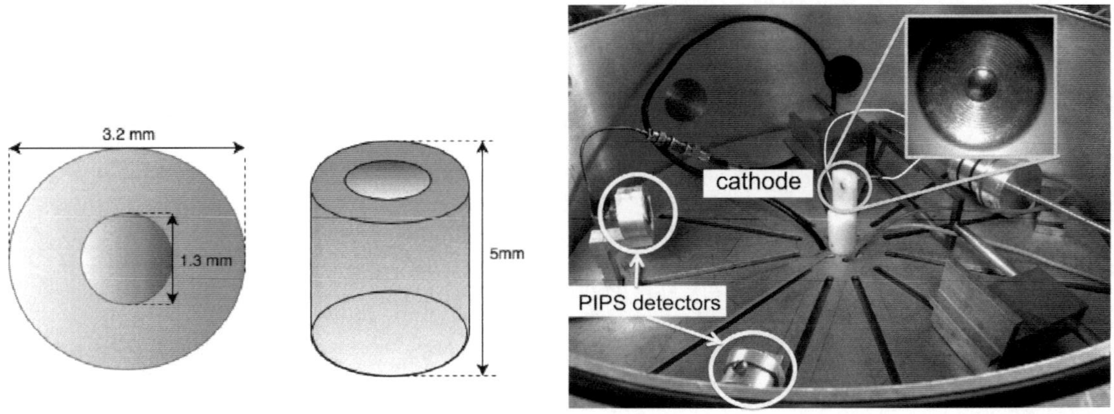

Figure 2. Diagram of a graphite-filled aluminium cathode (left); Set-up configuration for RBS(right)

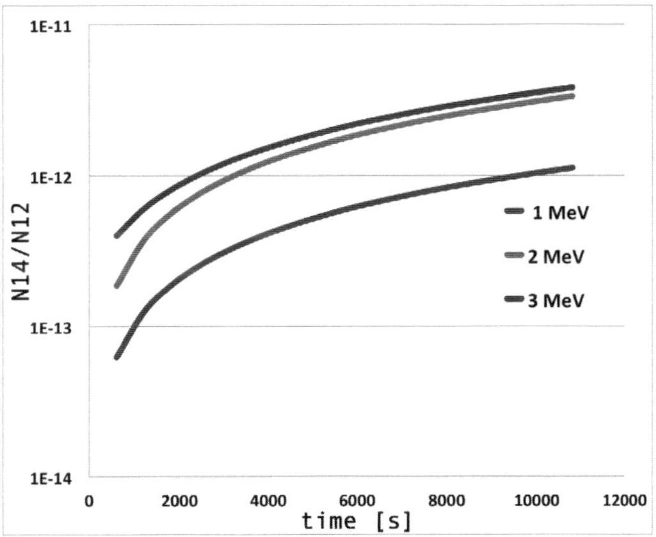

Figure 3. Plot of N_{14}/N_{12} vs. time for 1 MeV (blue), 2 MeV(green) and 3 MeV (red) for 10 hours of irradiation.

The variation with time of the ^{12}C, ^{13}C and ^{14}C concentrations is described by the equations below:

$$\frac{dN_{12}}{dt} = -\, N_{12}\,\Sigma_{12}\,\phi \quad , \tag{1}$$

$$\frac{dN_{13}}{dt} = N_{12}\,\sigma_{13}\,\phi \;-\; N_{13}\Sigma_{13}\phi \quad , \tag{2}$$

$$\frac{dN_{14}}{dt} = N_{13}\,\sigma_{14}\,\phi \;-\; N_{14}\Sigma_{14}\phi \quad . \tag{3}$$

where $N_{(12,13,14)}$ correspond to the concentrations of ^{12}C, ^{13}C, and ^{14}C at a given time t, $\Sigma_{12,13,14}$ are the sums of the cross sections of the reactions that consume ^{12,13,14}C ((d,p),(d,n),(d,γ) and (d,α)), $\sigma_{13,14}$ are the cross sections of the nuclear reactions that produce ^{13}C and ^{14}C, and

IOP Conf. Series: Journal of Physics: Conf. Series **876** (2017) 012017 doi:10.1088/1742-6596/876/1/012017

$\phi \approx 6.4 \times 10^{13}$ particles/s $\cdot$ cm^2 is the flux of deuterons impinging on the graphite surface of the cathode (corresponding typically to a 500 nA deuterium beam after collimation).

In Figure 4, the excitation energy is shown for the possible reaction channels, which can be determined initially by the laws of conservation of energy and changing from the laboratory system to the center-of-mass reference frame to obtain:

$$K_{thr(c.m.)} = -Q\left(1 + \frac{m_p}{m_t}\right) \tag{4}$$

where $K_{thr(c.m.)}$ is the threshold energy of the projectile, Q the $Q - value$ of the reaction channel, m_p the mass of the projectile and m_t the mass of the target.

Since natural graphite is a mixture of ^{12}C and ^{13}C, both ^{12}C(d , p) and ^{13}C(d , p) nuclear reactions take place. Therefore, we consider the reaction channels of each.

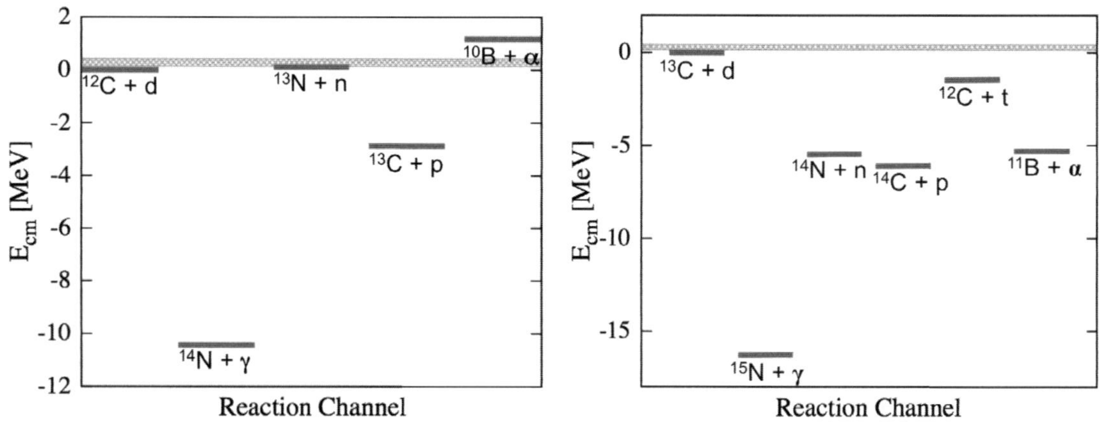

Figure 4. Excitation energy for the reaction channels of ^{12}C(d , p) and ^{13}C(d , p) in the center-of-mass reference frame. The gray band illustrates the energy region in which our study will take place.

3. Conclusions

A campaign of irradiation of cathodes containing graphite with a deuterium beam is planned to be performed at the 5.5 MV Van de Graaff accelerator in April, 2017, when we expect to be able to study the cross section for the ^{13}C(d, p) reaction at different beam energies between 1 and 3 MeV in the laboratory frame. The development of the AMS technique to measure cross sections of nuclear reactions is important since there is a large number of cross sections of astrophysical interest that could be studied with this method. Additionally, developing this technique would allow us to take advantage of the infrastructure that we have in Mexico to conduct studies that are in the frontier of physics.

Acknowledgments

The authors of this work are indebted to the Van de Graaff accelerator and LEMA technicians: G. Rayo and C. Valencia. And S. Martínez" This work is possible only thanks to the financial support from CONACYT 51600, 82692,123655, 271802 and DGAPAUNAM IN-118310, IN-112609, IN-105510, IN-115213, TA-100213, IN-103312 and IG-101016..

References

[1] He M *et al.*(2007)*Nucl. Instr. and Meth. in Phys. Res.* B **259** 629-31
[2] Wallner A *et al.* (2016) *Phys. R.* C **93**, 045803

IOP Conf. Series: Journal of Physics: Conf. Series **876** (2017) 012017 doi:10.1088/1742-6596/876/1/012017

[3] Wallner A *et al.* (2008) *J. Phys.* G*: Nucl. Part. Phys.* **35** 014018
[4] Arazi A *et al.* (2006) *Phys. Rev.* C **74** 025802
[5] Kouldijs B, Valckx F and Endt P (1953)*Physica* **19** 1133-39
[6] Marion J and Weber G (1956) *Phys. Rev.* **103**, 167-71
[7] Maxson D and Sornborger A (1993) *Nuclear Instr. and Meth. in Phys. R.* B **79** 293-96
[8] Steigman G 2003 Primordial Nucleosynthesis astro-ph/0308511
[9] Iocco F et al. 2007 The path to metallicity: synthesis of CNO elements in standard BBN *Phys. Rev.* D Phys.Rev.D75:087304,2007
[10] Koning A and Rochman D (2014) TENDL-2014 *Nuclear Research and Consultancy Group (NRG) Petten, The Netherlands.*

IOP Conf. Series: Journal of Physics: Conf. Series **876** (2017) 012018 doi:10.1088/1742-6596/876/1/012018

Aluminum and Oxygen Isotopic Ratios in Meteorite Grains: a Puzzle Solved By Nuclear and Stellar Physics

Sara Palmerini

Dipartimento di Fisica e Geologia Universitá degli Studi di Perugia, Italy
I.N.F.N. sezione di Perugia, Italy

E-mail: sara.palmerini@pg.infn.it

Abstract. Low mass stars contribute to the chemical evolution of the Galaxy as well as more massive supernova progenitors. Indeed the limited amount of processed matter released into the interstellar medium by small objects is compensated by the large number of them. At the late stages of their evolution, stars with mass smaller than $3M_{\odot}$ undergo the Asymptotic Giant Branch phase, which has been found to be a unique site for synthesis of some nuclei heavier than Fe through slow neutron capture reactions. AGB nucleosynthesis is also characterized by H-burning coupled with mixing phenomena, which has been proved to account for anomalies in light element isotopic abundances from Li to Al observed in stellar spectra and meteorites. We present here the case of a large number of meteorite grains, whose isotopic composition offers a puzzles that only Nuclear and Stellar Physics coupled together might solve.

1. Introduction

In February 1969, a rare type of meteorite (a Type III carbonaceous chondrite) broke up in the atmosphere producing the falling of thousands of stones over a nearly 500-square-kilometer area around the village of Pueblito de Allende, in the Northern Mexico. The Allende meteorite turned out to contain some of the solids formed in the Early Solar System as well as dust formed in previous generation of stars. In particular the meteorite is notable for possessing abundant, large calcium-aluminium-rich inclusions (CAIs), which are among the oldest objects formed in our planetary system and report an isotopic composition distinct from the one of the Earth. Beside CAIs, other micrometer to submicrometer inclusions in Allende, called "presolar grains", bring precious pieces of information about stellar nucleosynthesis. Indeed their "isotopic anomalies" contain evidence for processes that occurred in stars older that the Sun, and are of particular importance in helping us to constrain their nucleosynthesis.

A consistent part of presolar grains (commonly called oxide grains) is made up of Al_2O_3 grains, which are classified in four groups according to the Oxygen isotopic mix they show [1]. Grains belonging to group 1 and 2 are characterized by values of $^{17}O/^{16}O$ and $^{18}O/^{16}O$ inconsistent with explosive nucleosynthesis scenarios and are then believed to form in evolved low mass stars, during the Red Giant Branch (RGB) phase or Asymptotic Giant Branch (AGB) one [2]. However, the high dilution in ^{18}O and the large abundance of ^{26}Mg found in a large part of group 2 grains can not be accounted for by the nucleosyntesis of those stars, unless to invoke the existence of non-convective transport mechanisms coupled with H-burning ([3] and references therein).

Firstly, [4] suggested that the oxygen isotopic abundances of Al_2O_3 grains could hint to their formation in low mass red giant stars ($M \leq 3M_\odot$) where a deep and not convective mixing mechanism was at play. These authors developed a streams "conveyor belt" model able to enrich the stellar surface with fresh products of the CNO cycle by transporting matter downward, from the inner border of the stellar envelope to the regions where H-burning occurs, and upward, in the opposite direction. The work by [4] and the subsequent update were partially successful in reproducing the ^{18}O dilution shown by presolar grains, but part of group 1 and 2 oxide particles lays in regions of the ^{17}O/^{16}O versus ^{18}O/^{16}O diagram that were inaccessible to nucleosynthesis models (even considering deep mixing phenomena).

The ^{26}Mg excess found in presolar grains hints to their formation in ^{26}Al rich environments, for this reason [5] indicated as their stellar progenitors late AGB stars where the H-burning takes place at temperatures higher than those typical of the previous RGB stage and where a contribution to ^{26}Al production comes also from slow neutron capture process. Nevertheless, the ^{26}Al/^{27}Al ratio shown by several grains is higher than the level predicted by the standard AGB models (not including deep mixing phenomena). Dealing with non convective mixing contribution to AGB nucleosynthesis, [6] clarified that because of the long life of ^{26}Al ($\sim 7.17 \cdot 10^5$ years) its abundance is strongly related with the depth of mixing penetration while it does not significantly depend on the mixing velocity nor on the mixing rate. Indeed the isotope is produced by the H-burning shell via the ^{25}Mg(p,γ)^{26}Al reaction, which efficiently burns just at relatively high temperature. However, reproducing values of $^{26}\mathrm{Al}/^{27}\mathrm{Al} \geq 0.01$ requires to push the mixing into layers of the H-burning shell so deep and hot that the process will affect the stellar energy balance with relevant luminosity feedback.

The parametric models of the quoted authors were essentially based on a fine tuning of two free parameters: the amount of material mixed into the stellar envelope and the maximum penetration of the mixing. The first one depends in its turn both on the velocities of the upward and downward streams of matters, and on the fraction of the area, at fixed radius, occupied by the "conveyor-belt". Understanding the physical causes of extra-mixing would be a big step forward in our knowledge of stellar physics and it would provide a model for deep mixing whose characteristics will be fixed by the physical conditions of the stellar environments and no longer set by the fine tuning of free parameters. In this paper we will present the case of a mixing mechanism induced by stellar magnetic fields.

2. Nuclear physics inputs and calculations

The puzzle of the oxygen isotopic mix of group 1 and 2 oxide grains was solved in the last years thanks to a series of nuclear physics measurements, which allowed (i) to state that these grains can not be formed in massive AGB stars [7] and (ii) to obtein a full match between the dust oxygen abundances and the predictions of deep mixing models for low mass red giants [8, 9](and references therein). In particular, RGB stars with mass from 1 to $2M_\odot$ and solar composition were proved to be progenitors of group 1 grains thanks to the ^{14}N(p,γ)^{16}O cross section measurements by [8] as well as AGB stars with mass smaller than $1.5M_\odot$, experienced an efficient mixing, were shown to account for the Oxygen isotopic ratios of group 2 grains because of the determination of the low energy cross sections of proton-induced reactions on $A = 17$ and $A = 18$ oxygen isotopes via the Trojan Horse Method (THM, [10, 11]).

The THM overcomes enhancement effects due to electron screening and allows to measure reaction cross section between charged particles in the energy range typical of H-burning in low mass red giants (a few tens of keV). For this reason it was employed to extract the strength of the 65 keV resonance in the ^{17}O(p,α)^{14}N reaction [12], the strengths of the 20 keV one in the ^{18}O(p,α)^{15}N reaction and the contribution to its rate of the tail of the 656 keV broad resonance

[13]. Moreover, the strength of the 65 keV resonance in the ^{17}O(p,α)^{14}N reaction, measured by means of the THM, was used to renormalize the corresponding resonance strength in the ^{17}O + p radiative capture channel. As results, more accurate reaction rates for the ^{18}O(p,α)^{15}N , ^{17}O(p,α)^{14}N, and ^{17}O(p,γ)^{18}F processes were deduced [9, 14] and later introduced into the models for proton-capture nucleosynthesis coupled with mixing episodes presented by [15].

By employing those 3 reaction rates in an updated version of the "conveyor-belt" mixing model by [4] and [6], [9] succeed in reproducing the whole range of oxygen isotopic mix values measured in group 2 grains, indeed nucleosynthesis model curves reach the "forbidden" area of the ^{17}O/^{16}O vs ^{18}O/^{16}O diagram.

This sensitivity of the resulting oxygen isotopic mix to reaction rates adopted in calculations, and in particular of those of the ^{17}O + p process, is due to the fact that when an efficient deep mixing is at play the stellar envelope composition closely resemble the one of CNO equilibrium (at a temperature T, typical of the deepest layer reached by the streams of matter) which on its turn is determined by the equation

$$\frac{dY(^{17}O)}{dt} = Y(^{16}O)Y(H)R_{^{16}O(p,\gamma)} - Y(^{17}O)Y(H)(R_{^{17}O(p,\alpha)} + R_{^{17}O(p,\gamma)}) \quad (1)$$

where $Y(i)$ means the abundance in number of the ith isotope and any reaction rate R is temperature and density dependent. From the previous equation, the equilibrium value of the ^{17}O/^{16}O ratio turns out to be

$$\frac{Y(^{17}O)}{Y(^{16}O)} = \frac{R_{^{16}O(p,\gamma)}}{R_{^{17}O(p,\alpha)} + R_{^{17}O(p,\gamma)}} \quad (2)$$

One can notice that equilibrium values of the ^{17}O/^{16}O isotopic ratio do not depend on the initial isotopic abundances but are instead determined by the values of the reaction rates. As a consequence, in [9]'s calculations the ^{17}O/^{16}O ratio result larger because of THM cross sections of the ^{17}O + p reactions are smaller (see Figure 3, 4 and 5 in [9]) than those reported in [8].

Despite the nice agreement between the theoretical predictions and the dust contents in ^{17}O and ^{18}O and the success of the parametric extra-mixing model by [15, 9] in explaining the isotopic ratios of CNO elements, as observed in stellar photospheres [16], a big challenge about the ^{26}Mg excess found in presolar grains (hinting to their formation in a ^{26}Al-rich environments) remains. Especially after that a possible solution coming from nuclear physics was excluded by the measurement of the ^{25}Mg(p,γ)^{26}Al cross section published in [17] and by the recent one of the ^{26}Al(p,γ)^{27}Mg by [18].

3. Novelties from stellar physic

In the last years [19] demonstrated that the full MHD[1] equations can be solved analytically in an exact way when applied to the radiative layers of low mass evolved stars. This fact allows to build a mixing model where the mass transport is driven by the buoyancy of magnetized structures. Such an idea derives from an application to a later evolutionary phase of the dynamo model developed by Parker for the Sun [20]. Recently, [21] have verified that the physical conditions of the H-shell of RGB and AGB stars, with mass smaller than 2.5M$_\odot$, allow for the solution of MHD equations and have presented a non-parametric model for a mass circulation mechanism by applying the analytical solutions found by [19] to the stellar region of interest. In this framework, the mixing velocity is the radial component of the one by which magnetized structures (or flux tubes) expand and has the form:

$$v_r = v_0(\frac{r}{r_0})^{-(k+1)} \quad (3)$$

[1] magnetohydrodynamic

where r is the position along the stellar radius, while k is a number smaller than -1 dealing with the form of the plasma density distribution, which has to have the simple form $\rho \propto r^k$. In this approach the mixing depth is determined by k. Indeed the rising of magnetize structures starts from the deepest stellar layers whose physical conditions admit an analytical solution of MHD equations. [21] have shown that in low mass AGB stars those layers have the values of k ranging from -3.5 to -3.1 (the smaller is k the deeper is the mixing).

Moreover, in agreement with [21, 22] the mixing rate forced by magnetic buoyancy is given by

$$\dot{M} = 4\pi\rho_e r_e^2 v_e f_1 f_2 \tag{4}$$

where r_e is the stellar radius at the interphase between the radiative region and the convective envelope and v_e is the speed by which r_e is reached by raising materials. f_1 is the fractional area the flux tubes occupy when the stellar radius is r_e and f_2 is their filling factor. Extrapolating solar conditions to an evolved star, a field of a few 10^5 Gauss may be involved in magnetized structure occupying a fraction of $1-2\%$ of the whole area at the base of convective envelope ([19], and references therein), while f_2 is of about 1/100 of the section of the magnetized structures (which are actually almost empty, with electrically-charged materials concentrated in thin current sheets, see e.g. [23]). Since v_e has to be smaller than the convective velocity in the bottom envelope layers (v_{conv}) to allow the macroturbolence to destroy magnetized structure and release the trapped mass into the surface, we adopted a value of $v_e \leq v_{conv}$, which for AGB stars with $1.2 - 1.5 M_\odot$ is of the order of 100 m/s. In the light of the above estimate frr f_1, f_2 and v_e, the $\dot{M}$ values obtained range between a 0.3 and $1.3 \cdot 10^{-6} M_\odot/yr$, which are in the range found previously by [15] (with their parametric model of mixing) to account for the oxygen isotopic composition of group 2 oxide grains and the $^{12}C/^{13}C$ isotopic ratio observed in the spectra of MS and S stars [24].

4. Results

We applied a model of mass circulation induced by stellar magnetic fields [19] to study the isotopic composition of oxide presolar grains. In particular we aimed to reproduce the smaller values of ^{18}O as long with the higher one of ^{26}Mg (from the decay of ^{26}Al) measured in dust of RGB and AGB origin.

We performed our calculations employing the THM estimates of the $^{17}O(p,\alpha)^{14}N$, $^{17}O(p,\gamma)^{18}F$ and $^{18}O(p,\alpha)^{15}N$ low-energy reaction rates and assuming flux tubes rising from layers where $k = -3.5, -3.4, -3.3, -3.2$ and -3.1, as suggested by [21]. Computation results are shown in figure 1 for a $1.2 M_\odot$ evolved star with solar composition. In panel A the grey squares along the dashed line report the oxygen isotopic abundances in the envelope of stars approaching the RGB phase with different mass (from 1 to $1.7 M_\odot$ as indicated by the labels). Such values, which result to be sensitive to $^{14}N(p, \gamma)^{15}O$ but not affected by the ^{17}O + p reaction rates employed in stellar model calculation, remain almost unchanged for the rest of stellar evolution if no mixing phenomenon occurs. Since a dilution of the $^{18}O/^{16}O$ abundance in the stellar envelope is needed to reproduce the values measured in group 2 grains (solid points in figure 1A), our MHD mixing model has been introduce in nucleosynthesis calculations for both the RGB and AGB phase and the effects on the temporal evolution of chemical abundances are shown by the downward red curves (the dark is the color, the smaller is the value of k, the deeper is the mixing).

The sequence of curves cover the whole area where group 2 oxide grains lie. By using the same set for the nuclear physics input, we get therefore a confirmation of what was found by [9]: the dust in exam can be explained by most efficient mixing cases applied to a $1\text{-}1.5 M_\odot$ stars; but these conclusions are now based on a deep-mixing mechanism rooted in a more physical approach, and based on necessary processes of plasma physics.

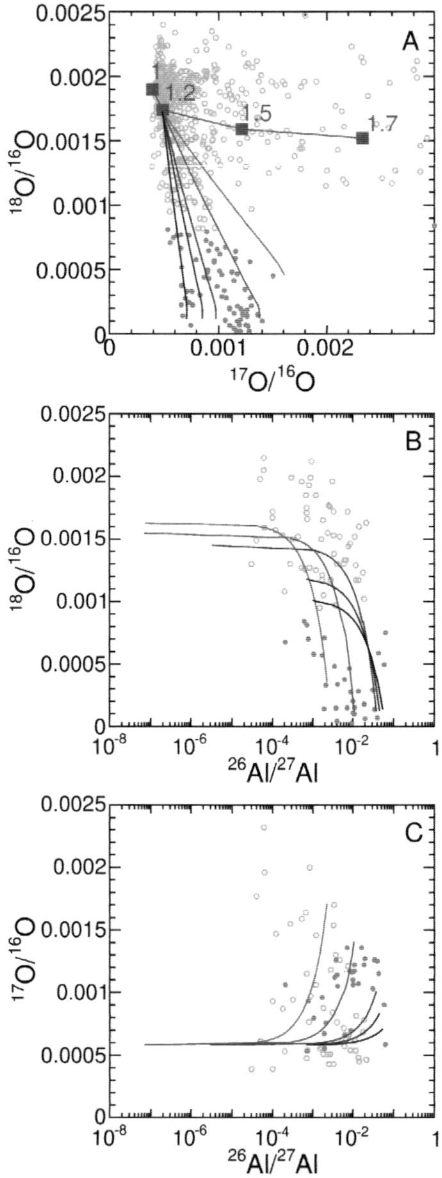

Figure 1. Panel A. Oxygen isotopic mix in the envelope of solar-metallicity stars approaching (dashed grey line). Full squares indicate the composition of stars with mass from 1 to $1.7M_{\odot}$, as indicated by the labels. Data points refer to measurements in presolar grains from [25]. We plot those of group 1 (open circles) and those of group 2 (filled circles). We do not report error bars in the figures, because they would confuse the plot, due to the large number of grains. In any case uncertainties are typically lower than 5-6% for all the grains of groups 1 and 2. The descending continuous lines refer to model results for deep mixing driven by magnetic buoyancy in a $1.2_{\odot}$ star for different values of k, during the RGB and the AGB phase. The smaller the value of k, the darker the curve ($k = -3.5, -3.4, -3.3, -3.2, -3.1$). Panel B and C. The evolution of the oxygen isotopic ratios $^{17}O/^{16}O$ and $^{18}O/^{16}O$, respectively, as functions of $^{26}Al/^{27}Al$ for the same models and grains reported in panel A,

In panels B and C, the evolution of the oxygen isotopic ratios is shown as a function of the ^{26}Al/^{27}Al ratio for the same models and grains discussed so far. The curves deal with the same magnetic mixing cases shown in panel A. We notice that the success of magnetically-induced mixing lays in accounting for the range of values of ^{26}Al/^{27}Al up to 0.1, even adopting the already-quoted rate by [17] for ^{26}Al production and the one by [18] for its destruction. We recall that previous parametric models could not, instead, explain the whole range of ^{26}Al, as measured in presolar grains, see e.g. [15] and in particular their Figure 11.

The success of the MHD induce mixing in reproducing the whole isotopic composition of group 2 grains is due to the fact that in this model hot magnetized materials are pushed up into the envelope from below (driven by magnetic buoyancy) and the down-flow occurs as a consequence of mass conservations at the envelope border. In the parametric mixing described by [4], and later by [6], upflowing streams of materials are triggered (because of the mass conservation law) by the penetration of matters from the bottom of the stellar envelope.

In conclusion, the whole areas that group 2 oxide presolar grain occupy in the ^{17}O/^{16}O vs ^{18}O/^{16}O diagram, in the ^{17}O/^{16}O vs ^{26}Al/^{27}Al one and in the ^{18}O/^{16}O vs ^{26}Al/^{27}Al one are covered by curves of a physically-based model, for mixing in low mass evolved stars, employing a set of nuclear physics input measured ad-hoc by the THM in the energy range typical of AGB star nucleosynthesis.

4.1. Acknowledgments
S.P. acknowledges the support of Fondazione Cassa di Risparmio di Perugia.

5. References
[1] Nittler L R 2009 *Publications of the Astron. Soc. of Australia* **26** 271–277 (*Preprint* 0904.1388)
[2] Zinner E K 2005 *Meteorites, Comets and Planets: Treatise on Geochemistry, Volume 1. Edited by A. M. Davis. Executive Editors: H. D. Holland and K. K. Turekian. ISBN 0-08-044720-1. Published by Elsevier B. V., Amsterdam, The Netherlands, 2005, p.17* ed Davis A M, Holland H D and Turekian K K (Elsevier B) p 17
[3] Busso M, Palmerini S, Maiorca E, Cristallo S, Straniero O, Abia C, Gallino R and La Cognata M 2010 *The Astrophysical Journal Letters* **717** L47–L51 (*Preprint* 1005.3549)
[4] Wasserburg G J, Boothroyd A I and Sackmann I J 1995 *The Astrophysical Journal Letters* **447** L37
[5] Boothroyd A I, Sackmann I J and Wasserburg G J 1994 *The Astrophysical Journal Letters* **430** L77–L80
[6] Nollett K M, Busso M and Wasserburg G J 2003 *The Astrophysical Journal* **582** 1036–1058 (*Preprint* astro-ph/0211271)
[7] Iliadis C, Angulo C, Descouvemont P, Lugaro M and Mohr P 2008 *Physics Review C* **77** 045802 (*Preprint* 0803.2757)
[8] Adelberger E G, García A, Robertson R G H, Snover K A, Balantekin A B, Heeger K, Ramsey-Musolf M J, Bemmerer D, Junghans A, Bertulani C A, Chen J W, Costantini H, Prati P, Couder M, Uberseder E, Wiescher M, Cyburt R, Davids B, Freedman S J, Gai M, Gazit D, Gialanella L, Imbriani G, Greife U, Hass M, Haxton W C, Itahashi T, Kubodera K, Langanke K, Leitner D, Leitner M, Vetter P, Winslow L, Marcucci L E, Motobayashi T, Mukhamedzhanov A, Tribble R E, Nollett K M, Nunes F M, Park T S, Parker P D, Schiavilla R, Simpson E C, Spitaleri C, Strieder F, Trautvetter H P, Suemmerer K and Typel S 2011 *Reviews of Modern Physics* **83** 195–246 (*Preprint* 1004.2318)
[9] Palmerini S, Sergi M L, La Cognata M, Lamia L, Pizzone R G and Spitaleri C 2013 *The Astrophysical Journal* **764** 128
[10] Spitaleri C, Cherubini S, del Zoppo A, di Pietrob A, Figuerab P, Gulino M, Lattuadab M, Miljanić D, Musumarra A, Pellegriti M G, Pizzone R G, Rolfs C, Romano S, Tudisco S and Tumino A 2003 *Nuclear Physics A* **719** C99–C106
[11] Spitaleri C, Cherubini S, La Cognata M, Lamia L, Mukhamedzhanov A, Pizzone R G, Romano S, Sergi M L and Tumino A 2010 *Nuclear Physics A* **834** 639–642c
[12] Sergi M L, Spitaleri C, La Cognata M, Coc A, Mukhamedzhanov A, Burjan S V, Cherubini S, Crucillá V, Gulino M, Hammache F, Hons Z, Irgaziev B, Kiss G G, Kroha V, Lamia L, Pizzone R G, Puglia S M R, Rapisarda G G, Romano S, de Séréville N, Somorjai E, Tudisco S and Tumino A 2010 *Physics Review C* **82** 032801
[13] La Cognata M, Spitaleri C and Mukhamedzhanov A M 2010 *The Astrophysical Journal* **723** 1512–1522

IOP Conf. Series: Journal of Physics: Conf. Series **876** (2017) 012018 doi:10.1088/1742-6596/876/1/012018

[14] Sergi M L, Spitaleri C, La Cognata M, Lamia L, Pizzone R G, Rapisarda G G, Tang X D, Bucher B, Couder M, Davies P, deBoer R, Fang X, Lamm L, Ma C, Notani M, O'Brien S, Roberson D, Tan W, Wiescher M, Irgaziev B, Mukhamedzhanov A, Mrazek J and Kroha V 2015 *Physics Review C* **91** 065803
[15] Palmerini S, La Cognata M, Cristallo S and Busso M 2011 *The Astrophysical Journal* **729** 3 (*Preprint* 1011.3948)
[16] Hedrosa R P, Abia C, Busso M, Cristallo S, Domínguez I, Palmerini S, Plez B and Straniero O 2013 *The Astrophysical Journal Letters* **768** L11 (*Preprint* 1303.6418)
[17] Straniero O, Imbriani G, Strieder F, Bemmerer D, Broggini C, Caciolli A, Corvisiero P, Costantini H, Cristallo S, DiLeva A, Formicola A, Elekes Z, Fülöp Z, Gervino G, Guglielmetti A, Gustavino C, Gyürky G, Junker M, Lemut A, Limata B, Marta M, Mazzocchi C, Menegazzo R, Piersanti L, Prati P, Roca V, Rolfs C, Rossi Alvarez C, Somorjai E, Terrasi F and Trautvetter H P 2013 *The Astrophysical Journal* **763** 100 (*Preprint* 1211.6661)
[18] Pain S D, Bardayan D W, Blackmon J C, Brown S M, Chae K Y, Chipps K A, Cizewski J A, Jones K L, Kozub R L, Liang J F, Matei C, Matos M, Moazen B H, Nesaraja C D, Okołowicz J, O'Malley P D, Peters W A, Pittman S T, Płoszajczak M, Schmitt K T, Shriner J F, Shapira D, Smith M S, Stracener D W and Wilson G L 2015 *Physical Review Letters* **114** 212501
[19] Nucci M C and Busso M 2014 *The Astrophysical Journal* **787** 141 (*Preprint* 1404.2503)
[20] Parker E N 1960 *The Astrophysical Journal* **132** 821
[21] Palmerini S, Trippella O and Busso M 2017 *Montly Notices of the Royal Astronomical Society* (*Preprint* 1701.04445)
[22] Trippella O, Busso M, Palmerini S, Maiorca E and Nucci M C 2016 *The Astrophysical Journal* **818** 125 (*Preprint* 1512.06777)
[23] Hirayama T 1992 *Solar Physics* **137** 33–50
[24] Wallerstein G, Balick B, Alcolea J, Bujarrabal V and Vanture A D 2011 *Astronomy and Astrophysics* **535** A101
[25] Hynes K M and Gyngard F 2009 *Lunar and Planetary Science Conference* (*Lunar and Planetary Science Conference* vol 40) p 1198

IOP Conf. Series: Journal of Physics: Conf. Series **876** (2017) 012019 doi:10.1088/1742-6596/876/1/012019

Global Correlations for Low-Lying Collective 2^+ States

Z. Z. Qin[1], Y. Lei[2], S. Pittel[3] and R. Bijker[4]

[1] School of Science, Southwest University of Science and Technology, Mianyang 621900, China
[2] Key laboratory of neutron physics, Institute of Nuclear Physics and Chemistry, China Academy of Engineering Physics, Mianyang 621900, China
[3] Bartol Research Institute and Department of Physics and Astronomy, University of Delaware, Newark, Delaware 19716, USA
[4] Instituto de Ciencias Nucleares, Universidad Nacional Autónoma de México, A.P. 70-543, 04510 México, D.F., México

E-mail: pittel@bartol.udel.edu

Abstract. By using the triaxial rotor model and the anharmonic vibrator model with phonon mixing, we derive a global correlation between the quadrupole moments of the two lowest 2^+ states in collective nuclei that had previously been observed in experimental data across the periodic table. We then derive other electromagnetic properties for these two models of nuclear structure and compare them globally with experimental data. We find that both models are able to robustly describe the experimental data across the region of nuclei for which the models are applicable, including a large number that they have in common. We then show that there seems to exists a robust orthogonal transformation between these two models for realistic nuclear systems, suggesting that these two seemingly diverse descriptions of quadrupole collective phenomena seem to act in a similar model space and may therefore have a common origin.

1. Introduction

A large range of nuclei across the periodic table exhibit quadrupole collectivity. This includes nuclei that exhibit rotational, vibrational, triaxial and gamma-unstable features. Though these nuclei seem to have a wide variety of distinct properties, there nevertheless appear some global features common to all. Here we will focus on one such feature reported recently in an experimental survey by Allmond [1]. What Allmond found is that for a wide range of masses and for $R = E(4_1^+)/E(2_1^+)$ values ranging from 2 (the vibrational limit) to 10/3 (the rotational limit) the quadrupole moments of the lowest 2^+ states for most of the nuclei in which both are measured satisfy the property that $Q(2_1^+) \sim -Q(2_2^+)$. Allmond noted that this feature is reproduced theoretically for most nuclei by the IBM-1.

In this work, we will discuss two phenomenological models of nuclear structure that analytically reproduce this global $Q(2_1^+) \sim -Q(2_2^+)$ correlation. One is the Triaxial Rotor Model (TRM) [2] and the other is the Anharmonic Vibrator Model (AVM) [3, 4, 5] with phonon configuration mixing. As we will see, both not only incorporate the above quadrupole moment correlation but both also do an excellent job of reproducing collective properties of low-lying quadrupole states across the periodic table.

The outline of the presentation is as follows. In Section 2, we briefly review the key elements of the Triaxial Rotor Model and the Anharmonic Vibrator Model with phonon mixing, focussing on those features of the models needed for the analysis to follow. As we will see the quadrupole moment correlation noted above is analytically present in both models. Then in Section 3, we confront these two models with experimental data across several regions of the nuclear periodic table in which quadrupole collectivity is observed. As we will see both models are able to successfully reproduce the data. In Section 4, we discuss the possible relationship between these two apparently distinct collective models of nuclear structure by analyzing the orthogonal transformation that connects their bases for 2^+ states. In Section 5, we present further numerical support for the quadrupole moment correlation underlying this work. Finally in Section 6 we summarize the key conclusions of this work and describe some possible avenues for future investigation. Further details on the work described herein can be found in Ref. [6].

2. Two phenomenological models of nuclear collective motion

2.1. The TRM

The Triaxial Rotor Model (TRM) is a phenomenological model of nuclear structure. The model contains a total of five parameters and has been used extensively to describe experimental data on E2 collectivity across several regions of the periodic table [7, 8, 9, 10].

Rather than describe the model in any detail, we refer the reader to a nice article where those details can be found [11]. We will, however, briefly summarize the key components of the formalism, as needed for our study.

The Hamiltonian matrices for 2^+ and 4^+ states in the model are written schematically as:

$$H^{2^+}_{\rm TRM} = \begin{pmatrix} 6A & 4\sqrt{3}G \\ 4\sqrt{3}G & 6A+4F \end{pmatrix} ,$$

$$H^{4^+}_{\rm TRM} = \begin{pmatrix} 20A & 12\sqrt{5}G & 0 \\ 12\sqrt{5}G & 20A+4F & 4\sqrt{7}G \\ 0 & 4\sqrt{7}G & 20A+16F \end{pmatrix} , \qquad (1)$$

where A, G and F are Hamiltonian parameters related to the three components of the inertia tensor.

The 2^+ states are orthogonal combinations of the $K=0$ and $K=2$ basis configurations, according to

$$|2^+_1\rangle = \cos\Gamma|K=0\rangle - \sin\Gamma|K=2\rangle \; ,$$
$$|2^+_2\rangle = \sin\Gamma|K=0\rangle + \cos\Gamma|K=2\rangle \; . \qquad (2)$$

Note that K is the projection of the angular momentum with respect to the intrinsic coordinate system, and $\tan 2\Gamma = 2\sqrt{3}G/F$ specifies the mixing of these basis configurations.

The E2 properties of the lowest 2^+ states can be expressed fairly simply as

$$\mathrm{B(E2,}2^+_1 \to 0^+_1) = \frac{Q_0^2}{16\pi}\cos^2(\gamma+\Gamma) \; ,$$
$$\mathrm{B(E2,}2^+_2 \to 0^+_1) = \frac{Q_0^2}{16\pi}\sin^2(\gamma+\Gamma) \; ,$$
$$\mathrm{B(E2,}2^+_2 \to 2^+_1) = \frac{5Q_0^2}{56\pi}\sin^2(\gamma-2\Gamma) \; ,$$
$$Q(2^+_1) = -\frac{2}{7}\,Q_0\cos(\gamma-2\Gamma) = -Q(2^+_2) \; . \qquad (3)$$

IOP Conf. Series: Journal of Physics: Conf. Series **876** (2017) 012019 doi:10.1088/1742-6596/876/1/012019

Here Q_0 is the static quadrupole moment, and γ is a parameter related to the nuclear quadrupole deformation. Adding these to the three independent inertia parameters gives a total of five parameters in the model.

Note from the last line of eq.(3) that a $Q(2_1^+) = -Q(2_2^+)$ correlation is unconditionally conserved in the TRM.

2.2. The AHV

In the AHV description of nuclei [4], the model space of 2^+ states is constructed from one- and two-phonon excitations of the phonon vacuum $|0\rangle$ (the 0^+ ground state of the system). We will denote them as $|1\rangle$ and $|2\rangle$, respectively.

In this model space, the Hamiltonian matrix for 2^+ states is given by

$$H_{\rm AHV}^{2^+} = \begin{pmatrix} \hbar\omega & \lambda \\ \lambda & 2\hbar\omega \end{pmatrix} , \tag{4}$$

where $\hbar\omega$ is the phonon energy and λ defines the mixing between one- and two-phonon configurations. The eigenstates of such a Hamiltonian are given by

$$\begin{aligned} |2_1^+\rangle &= a_1|1\rangle + a_2|2\rangle , \\ |2_2^+\rangle &= -a_2|1\rangle + a_1|2\rangle , \end{aligned} \tag{5}$$

where $a_1^2 + a_2^2 = 1$.

The E2 operator in the model is $\hat{Q} = \chi(b^\dagger + \tilde{b})$, in terms of phonon creation and annihilation operators and a free parameter χ. Quadrupole properties in the model can be summarized as

$$\begin{aligned} \mathrm{B(E2,}2_1^+ \to 0_1^+) &= \frac{\chi^2 a_1^2}{5}\langle 0||\tilde{b}||1\rangle^2 , \\ \mathrm{B(E2,}2_2^+ \to 0_1^+) &= \frac{\chi^2 a_2^2}{5}\langle 0||\tilde{b}||1\rangle^2 , \\ \mathrm{B(E2,}2_2^+ \to 2_1^+) &= \frac{\chi^2 (a_1^2 - a_2^2)^2}{5}\langle 1||\tilde{b}||2\rangle^2 , \\ Q(2_1^+) &= \frac{8\chi a_1 a_2}{5}\sqrt{\frac{2\pi}{7}}\langle 1||\tilde{b}||2\rangle = -Q(2_2^+) . \end{aligned} \tag{6}$$

Here too we find the desired quadrupole moment correlation, independent of the spectral behavior. However, we should emphasize that this only arises when we ignore multi-phonon states. The fact that data suggests that the quadrupole moment correlation exists is an indication that multi-phonon mixing should not be too important.

3. Systematic Comparison with Data

3.1. Spectral properties

Diagonalization of Eq. (1) for given values of A, F and G provides excitation energies of 2_1^+, 2_2^+ and 4_1^+ states in the TRM approach. We refer to those nuclei for which the approach leads to a fit to the data [12] as TRM-solvable nuclei and present in Fig. 1 the $R = E(4_1^+)/E(2_1^+)$ distribution of these 203 nuclei. As is evident from the figure, the distribution spreads over the whole $R > 2$ region, indicating that the TRM can indeed provide non-rotor-like spectra, while still maintaining the rotational quadrupole moment correlation. We also note that there are two peaks in the distribution. The peak near $R = 10/3$ peak corresponds to rotational nuclei with nearly perfect axially-symmetric deformation. The peak near $R = 5/2$ corresponds to γ-unstable nuclei, or nuclei near the $SO(6)$ limit of the IBM [13].

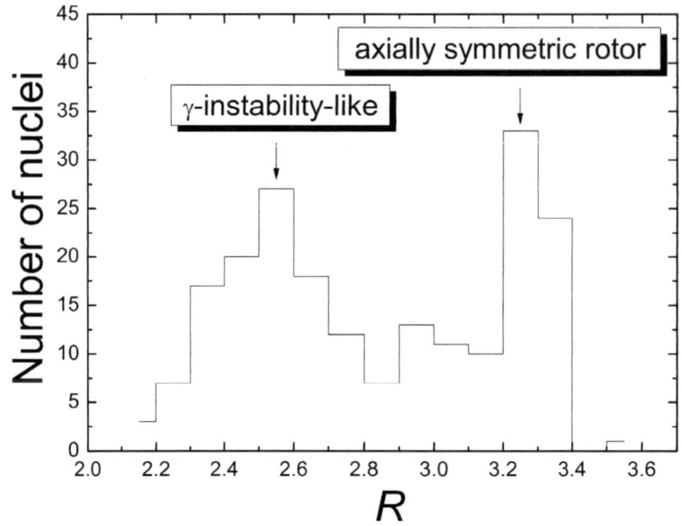

Figure 1. $R = E(4_1^+)/E(2_1^+)$ distribution of the 203 TRM-solvable nuclei in the ENSDF. Peaks for γ-instability ($R \simeq 5/2$) and an axially-symmetric rotor ($R \simeq 10/3$) are highlighted.

The spectral structure of nuclei within the AHV approach is simpler than for the TRM. Since the 2_1^+ and 2_2^+ states arise from mixing one- and two-phonon configurations, the energy of the lowest 2^+ state must be below the one-phonon excitation energy, $\hbar\omega$, whereas the energy of the second 2^+ state must be above $2\hbar\omega$. Fig. 2 shows all available experimental data on nuclei with $E(2_2^+)$ versus $E(2_1^+)$ in the range $R = 2.05 \sim 3.15$, nuclei that were assigned to the AHV by Casten *et al.*. This ensemble includes 177 nuclei in the ENSDF. Clearly most such nuclei sit above the $E(2_2^+) = 2E(2_1^+)$ line. This suggests that the ensemble of $R = 2.05 \sim 3.15$ nuclei is indeed a reasonable sample of AHV-applicable nuclei, in agreement with the classification of Casten *et al.* [14].

3.2. E2 collectivity

We now focus on correlations between E2 transition rates and the quadrupole moments of the two lowest 2^+ states in the two models. From earlier formulae presented for TRM we can obtain the following approximate relation between the BE2 values and quadrupole moments of 2^+ states:

$$\mathrm{B(E2,}\ 2_1^+ \to 0_1^+) + \mathrm{B(E2,}\ 2_2^+ \to 0_1^+) \simeq 0.7\ \mathrm{B(E2,}\ 2_2^+ \to 2_1^+) + 0.244\ Q^2(2_1^+)\ . \tag{7}$$

For the AHV, we can derive an analogous formula:

$$\mathrm{B(E2,}\ 2_1^+ \to 0_1^+) + \mathrm{B(E2,}\ 2_2^+ \to 0_1^+) \simeq 0.5\ \mathrm{B(E2,}\ 2_2^+ \to 2_1^+) + 0.174\ Q^2(2_1^+)\ . \tag{8}$$

Both are in the general form of Kumar-Cline sum rules and can be more generally written as

$$\mathrm{B(E2,}\ 2_1^+ \to 0_1^+) + \mathrm{B(E2,}\ 2_2^+ \to 0_1^+) = c_1\ \mathrm{B(E2,}\ 2_2^+ \to 2_1^+) + c_2\ Q^2(2_1^+)\ , \tag{9}$$

where c_1 and c_2 are free variables.

We perform a multiple linear fitting of this formula (9) to all available experimental data from the ENSDF with c_1 and c_2 as fitting parameters. A total of 78 nuclei are considered in the fit of the left side to the right side, with the correlation coefficient $r = 0.980$ that emerges

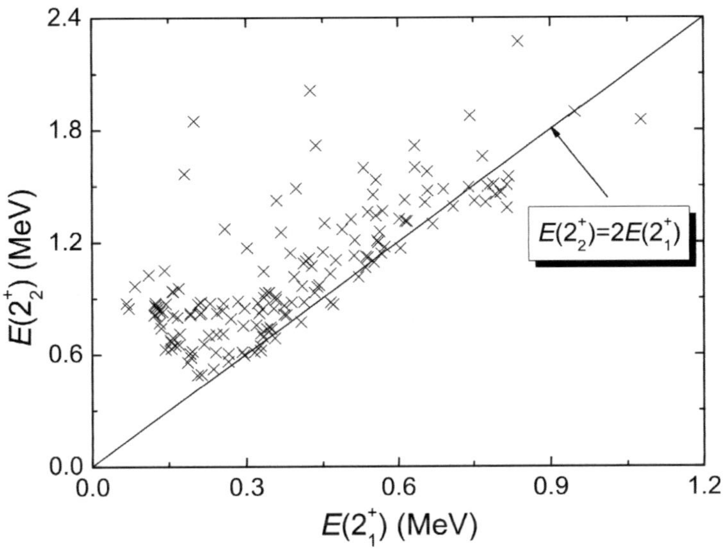

Figure 2. Experimental $E(2_2^+)$ versus $E(2_1^+)$ for the 177 nuclei with $R = 2.05 \sim 3.15$ in the ENSDF. The $E(2_2^+) = 2E(2_1^+)$ line is highlighted.

being very close to 1. This confirms that B(E2) values between ground states and low-lying 2^+ states are highly correlated with $Q(2_1^+)$ in the ENSDF, as expected for the two models under consideration. The best-fit results are $c_1 = 0.479$ and $c_2 = 0.188$, substantially closer to the AHV values.

Using Eqs. (7) for the TRM and (8) for the AHV, we estimate the magnitudes of the quadrupole moments $Q(2_1^+)$ from the experimentally available B(E2) values. In Fig. 3, we plot the $|Q(2_1^+)|$ values that emerge in comparison with the associated experimental values. For both models, the data points of both models scatter fairly closely around the diagonal line, confirming that both the TRM and AHV approaches provide meaningful global descriptions of low-lying E2 collectivity.

3.3. Other collective observables

We have also analyzed magnetic moments of the two lowest 2^+ states within the TRM and AHV frameworks. We summarize here the key results, with further details available in Ref. [6].

- In the AHV approach with mixing between one- and two-phonon configuration only, another exato ct correlation emerges, namely that $\mu(2_1^+) = \mu(2_2^+)$. Experimentally, this correlation is borne out extremely well for most of the AHV-applicable nuclei.
- In the TRM approach, the $\mu(2_1^+) = \mu(2_2^+)$ correlation that is present in experimental data only emerges if we assume that $g_R = g_K$. When the accepted g-factors and the parameters emerging from the fits to energies described earlier are used, one typically finds $\mu(2_1^+) \approx 0.8\mu(2_2^+)$ in the TRM.
- Thus, as for quadrupole moments, magnetic moments seem to be somewhat better described by the AHV approach than the TRM approach.

4. What next?

We have seen that both the TRM and the AHV provide a fairly robust description of excitation energies, E2 collective features and magnetic moments, especially for collective 2^+ states.

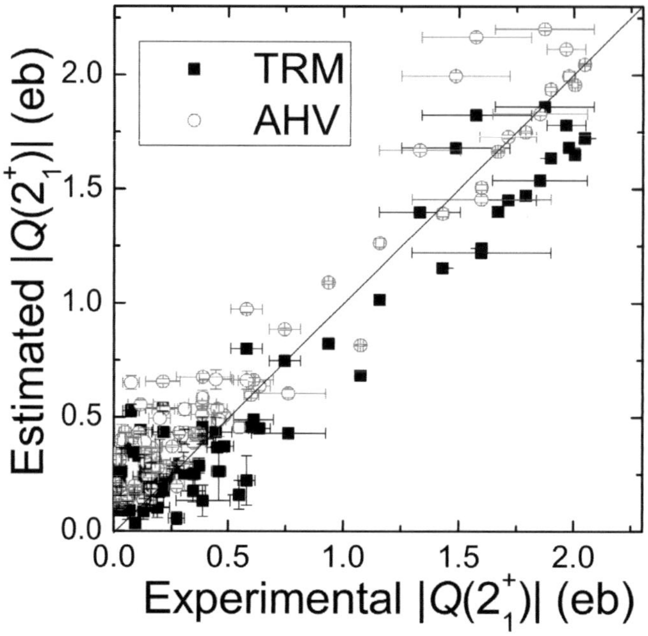

Figure 3. Plot of the estimated $|Q(2_1^+)|$ values based on the use of experimental B(E2) values according to Eqs. (7) and (8), against the experimental data, for the 78 nuclei with available data in the ENSDF. The diagonal line is a measure of the quality of the estimates.

Furthermore, in both cases, the 2^+ states are described in terms of a two-state basis. Since both models do a reasonable job of describing the same set of properties in terms of 2x2 matrices, there should be an orthogonal transformation for each such nucleus,

$$U = \begin{bmatrix} \cos\theta & \sin\theta \\ -\sin\theta & \cos\theta \end{bmatrix}, \tag{10}$$

such that

$$H^{2+}_{\mathrm{TRM}} = U H^{2+}_{\mathrm{AHV}} U^{\mathrm{T}}. \tag{11}$$

Is there an intrinsic relation between the TRM and the AHV models for realistic nuclei? If so, it would seem that there should be a unique orthogonal transformation that relates the two bases, namely the K basis of the TRM and the phonon basis of the AHV, *viz.*

$$\begin{bmatrix} K=0 \\ K=2 \end{bmatrix} = U \begin{bmatrix} |1\rangle \\ |2\rangle \end{bmatrix}. \tag{12}$$

To see whether the U matrix, i.e. the θ angle, is globally unique or robust for realistic nuclei, we calculate the distribution of θ values. We do this in such a way as to remove any bias due to the fact that the θ value for a given nucleus should most likely be correlated with the R value for that same nucleus.

We will not discuss here the details of how we do this, referring the reader to [6]. We should note, however, that we only get info on $|\theta|$, since θ and $-\theta$ define the same transformation.

The resulting distribution of $|\theta|$ values, denoted $P(|\theta|)$, is shown in Fig. 4 for the 203 nuclei in the ENSDF that are TRM- and AHV-solvable. What we see is that there is a single dominant

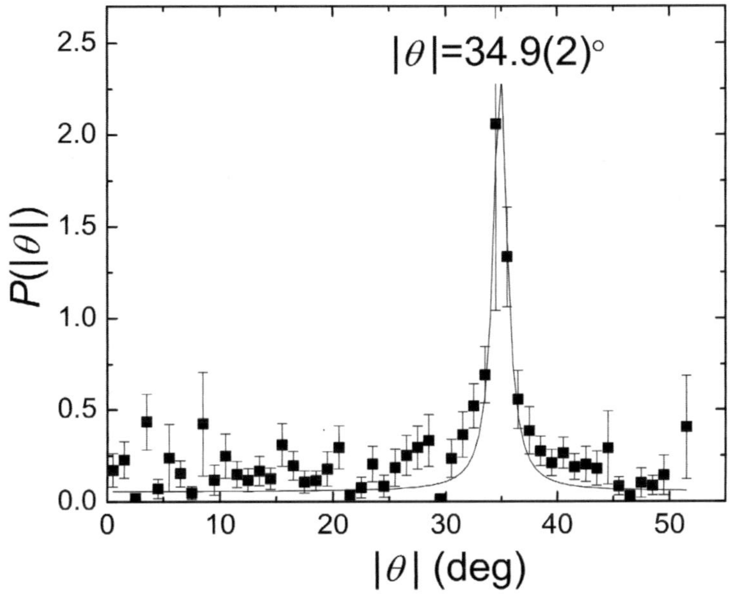

Figure 4. R-normalized $P(|\theta|)$ distribution (square points). The error bar represents the statistical error. The peak fit (solid line) has its center at $|\theta| = 34.9(2)^\circ$, as highlighted. This figure uses the same statistical ensemble as in Fig. 1.

peak, centered near $|\theta| = 34.9^\circ$, suggesting that the bases used in the two sets of analyses (for 2^+ states) seem to a good approximation be the same.

This raises the question of whether the underlying physics of quadrupole collective motion may have a deeper origin, with the TRM and AHV descriptions contained within a richer model of quadrupole collectivity.

As an illustration of what we have in mind, we have studied the behavior of the IBM-1 [13], but with just two bosons. As a reminder, the IBM-1 is a model of s ($L = 0$) and d ($L = 2$) bosons, whereby it is governed by a U(6) symmetry. The model has three dynamical symmetry limits, each describing a distinct types of quadrupole collectivity. For a perfect harmonic vibrator, the U(5) symmetry limit applies, with one-phonon and two-phonon states involving one and two d bosons, respectively. If we mix these states, we obtain the AHV model described earlier. If the mixing is driven by an SU(3) Hamiltonian, however, we find that a *unique* linear transformation of such phonon states provides two rotational 2^+ states, as defined in Eq. (10). We have calculated the transformation angle between the states of this basis and the lowest two 2^+ states of the SU(3) basis numerically, and find a rotation angle of $\theta = 28.1^\circ$, not too far from the 35° we found that relates the TRM and AHV bases fairly robustly. Of course, the TRM does not derive from the SU(3) limit of the IBM-1, which is an axially symmetric rotor, so that we cannot meaningfully compare the two angles that emerged. Nevertheless, this illustrates how two distinct models of nuclear collectivity can be a reflection of a richer model that contains both (and perhaps others).

5. The global quadrupole moment correlation in the IBM

In his article in which the global correlation $Q(2_1^+) = -Q(2_2^+)$ was first introduced, Allmond [1] pointed out that the correlation seems to be reproduced fairly generally in the IBM-1 in the context of a consistent-Q treatment. Numerical results were, however, not presented. Here we briefly expand on those considerations by presenting numerical results under a variety of different scenarios, now using the IBM-2.

In Table 1, we present results at or near a variety of symmetry limits. The results denoted SU(3), SU(5) and SU(3)* refer to symmetry limits. Both SU(3) and SU(5) refer to symmetric scenarios and could be obtained alternatively using an IBM-1 code. The SU(3)* results, which refers to a pure triaxial rotor [15], *required* a full IBM-2 treatment. In the exact O(6) limit, for which $\chi_\pi = \chi_\nu = 0$, the quadrupole moments are identically zero. Thus we consider a slight breaking of the symmetry (denoted by δ) by including a small amount of either SU(3) mixing or SU(3)* mixing. We do this by setting $\chi_\pi = \chi_\nu = 0.10$ (for SU(3)mixing) and $\chi_\pi = 0.10$, $\chi_\nu = -0.10$ (for SU(3)* mixing).

Table 1. Selected results for the ratio of quadrupole moments of the lowest two 2^+ states in the IBM-2 under a variety of conditions. All results are for $N_\pi = 3$ and $N_\nu = 5$ bosons.

	$Q(2_1^+)/Q(2_2^+)$
$SU(3)$	−1.20
$SU^*(3)$	−1.00
$U(5)$	−2.28
$SO(6) + \delta(SU(3))$	−1.04
$SO(6) + \delta(SU^*(3))$	−1.04

We should note that we have also carried out calculations for the slightly larger boson numbers $N_\pi = 5$ and $N_\nu = 7$ to see whether the boson number dramatically impacted the results. Though the precise numerical results changed slightly, the same general features emerged.

In Table 2, we explore the transition between vibrational and deformed nuclei, through the use of a parametrized IBM-2 Hamiltonian

$$H = (1-\xi)(n_{d_\pi} + n_{d_\nu}) - \xi \frac{N_\pi + N_\nu}{N_\pi N_\nu} Q_\pi \cdot Q_\nu \ , \tag{13}$$

with ξ varying from 0 to 1. Note that when $\xi = 1$, the Hamiltonian produces rotational motion, which smoothly evolves to vibrational motion as ξ is decreased to 0. At $\xi \approx 0.2$, a shape phase transition takes place and the spectrum involves significant level crossings. Thus, at this point we show the ratio of $Q(2_1^+)/Q(2_4^+)$ to reflect the 2^+ states with the same structure as prior to the transition. As is evident from the table, prior to the shape transition the ratio of quadrupole moments is roughly $-7/3$, as is typical of vibrational nuclei. After the shape transition, the ratio quickly falls to a value typical of rotational nuclei.

All results are compatible with $Q(2_1^+)$ and $Q(2_2^+)$ having opposite signs and roughly the same magnitudes (typically within a factor of roughly 2). However, as was also noted by Allmond, in the rotational scenario it is critical that the second 2^+ state not be part of a β band for this to emerge.

Table 2. Results for the ratio of quadrupole moments of the lowest two 2^+ states in the IBM-2 across the transition from vibrational to rotational nuclei, as produced by the Hamiltonian in Eq. 13. All results are for $N_\pi = 3$ and $N_\nu = 5$ bosons. At $\xi = 0.2$, the results denoted by a * refer to $Q(2_1^+)/Q(2_4^+)$.

ξ	$Q(2_1^+)/Q(2_2^+)$
0.0	−2.33
0.1	−2.30
0.2*	−2.31
0.3	−1.46
0.4	−1.27
0.5	−1.24
0.6	−1.23
0.7	−1.25
0.8	−1.20
0.9	−1.21
1.0	−1.21

6. Summary

In this article, we discussed two phenomenological models of nuclear structure, the TRM and the AHV, both of which have the feature that they analytically satisfy the property $Q(2_1^+) = -Q(2_2^+)$ that has been found globally for nuclei exhibiting quadrupole collectivity. We have shown that both of these models can describe systematic correlations between excitation energies and E2 collectivity, especially for the lowest two 2^+ states, across a wide range of the periodic table, and that they provide in general good overall agreement with the associated experimental data. We also find that the TRM and AHV Hamiltonian matrices can be connected by an orthogonal transformation that seems to be roughly the same for most nuclei. This suggests that the TRM and AHV, though seemingly very different models of nuclear collective behavior, seem to share the same model space for realistic nuclear systems, and perhaps are just two aspects of a more general model of quadrupole collective behavior. The Neutron-Proton Interacting Boson Model (IBM-2) seems a natural framework in which to explore this further.

Acknowledgments

This work was supported in part by the National Natural Science Foundation of China under Grant No. 11647059 and 11305151, the Research Fund for the Doctoral Program of the Southwest University of Science and Technology under Grant No. 14zx7102, and by PAPIIT-DGAPA, UNAM under Grant No. IN109017.

References

[1] J. M. Allmond, Phys. Rev. C **88** (2013) 041307.
[2] A. S. Davydov and G. F. Filippov, Nucl. Phys. **8**, 237 (1958).
[3] F. K. McGowan, R. L. Robinson, P. H. Stelson, and J. L. C. Ford Jr., Nucl. Phys. **66**, 97 (1965).
[4] T. Tamura and T. Udagawa, Phys. Rev. **150**, 783 (1966).
[5] D. R. Bes and G. G. Dussel, Nucl. Phys. A **135**, 1 (1969).
[6] Z. Z. Qin, Y. Lei and S. Pittel, Phys. Rev. C **94**, 024301 (2016).
[7] J. M. Allmond, R. Zaballa, A. M. Oros-Peusquens, W. D. Kulp, and J. L. Wood, Phys. Rev. C **78**, 014302 (2008).
[8] J. M. Allmond, J. L. Wood, and W. D. Kulp, Phys. Rev. C **81**, 051305 (2010).
[9] W. D. Kulp, *et al.*, Phys. Rev. C **73**, 014308 (2006).

[10] A. D. Ayangeakaa, *et al.*, Phys. Lett. B **754**, 254 (2016).
[11] J. L. Wood, A-M. Oros-Peusquens, R. Zaballa, J. M. Allmond, and W. D. Kulp, Phys. Rev. C**70** (2004) 024308.
[12] Evaluated Nuclear Structure Data File Retrieval, http://www.nndc.bnl.gov/ensdf/
[13] F. Iachello and A. Arima, *The Interacting Boson Model* (Cambridge University, Cambridge, UK, 1987).
[14] R. F. Casten, N. V. Zamfir, and D. S. Brenner, Phys. Rev. Lett. **71**, 227 (1993).
[15] A.E.L. Dieperink and R. Bijker, Phys. Lett. B **116**, 77 (1982).

IOP Conf. Series: Journal of Physics: Conf. Series **876** (2017) 012020 doi:10.1088/1742-6596/876/1/012020

The theory of hadronic parity violation

Matthias R Schindler

Department of Physics and Astronomy, University of South Carolina, Columbia, SC 29208, USA

E-mail: `mschindl@mailbox.sc.edu`

Abstract. Parity-violating interactions between nucleons are the manifestation of an interplay of strong and weak interactions between quarks in the nucleons. Compared to the dominant parity-conserving part, the parity-violating component of the nuclear force is typically suppressed by approximately 6 to 7 orders of magnitude or more. Due to the short range of the weak interactions, however, it provides a unique probe of the strong dynamics that confine quarks into nucleons. An ongoing experimental program is mapping out this weak component of the nuclear force in few-nucleon systems. I will discuss recent theoretical progress based on effective field theory methods to analyze and interpret hadronic parity violation in few-nucleon systems, with a particular focus on two- and three-nucleon systems.

1. Introduction

The forces between nucleons are the manifestation of interactions between the quarks confined in these nucleons. While nucleon interactions are dominated by strong and electromagnetic effects, the confined quarks are also subject to the weak interactions. This leads to a parity-violating (PV) component in the interactions between nucleons. At the low energies typical of few-nucleon experiments, the PV part is expected to be suppressed by 6 to 7 orders of magnitude compared to the parity-conserving (PC) component. Effects due to the highly suppressed PV interaction can be isolated by considering observables that would vanish if parity was conserved. Typically, these are pseudoscalar observables involving a spin and a momentum vector, such as longitudinal and angular asymmetries and induced polarizations. The weak interactions are well understood at the quark level; however, when viewed at the hadronic level they are intertwined with the nonperturbative strong effects that confine the quarks in hadrons. Because of their short range, weak interactions are sensitive to quark-quark correlations at very short distances. PV nucleon interactions can therefore be viewed as a unique probe of nonperturbative strong effects.

PV effects can be enhanced by several orders of magnitude in some complex nuclei, e.g., by the presence of closely-spaced energy levels with opposite parity (see, e.g., Ref. [1]). While this is of great advantage experimentally, the theoretical description of such complex systems in terms of nucleon-nucleon interactions is difficult and can lead to uncontrolled theoretical errors. With the development of high-intensity neutron and photon sources and improved control over systematics, the measurement of PV effects in few-nucleon systems with $A \leq 5$ has become feasible, and an experimental program is underway to map out the PV component of the nucleon interactions. In these light systems, the theoretical description in terms of two- and three-nucleon systems is much more feasible. For recent reviews, see, e.g., Refs. [2, 3].

Traditionally, PV nucleon-nucleon interactions have been most commonly described in terms of meson-exchange potentials with one PC and one PV meson-nucleon vertex. In particular, the formulation of Ref. [4] (often referred to as the DDH potential) has been used very frequently. More recently, effective field theory (EFT) methods, which have proven very successful in the PC sector, have been applied to PV observables. EFTs are model independent, treat PC and PV interactions as well as the coupling to external currents within a unified framework, and provide a method to estimate theoretical errors. The EFT analysis is performed at the hadronic level and does not directly relate the PV effects to interactions between standard model degrees of freedom. However, it provides a set of well-defined quantities with which any calculation performed at the quark level, such as potentially lattice QCD calculations, must be consistent.

2. Parity violation in pionless EFT

Many of the experiments in few-nucleon systems are performed at very low energies, e.g., using cold neutrons. At these energies, an EFT in which nucleons are the only dynamical degrees of freedom can be employed. In this so-called pionless EFT (EFT($\not\pi$)) pions and all other degrees of freedom are integrated out and interactions between nucleons are described by contact terms with an increasing number of derivatives. All short-distance details of the interactions are encoded in the low-energy constants (LECs) of the theory that accompany each operator in the Lagrangian. The power counting of the theory provides a systematic method to determine observables order-by-order in terms of a small expansion parameter given by a ratio of typical scales involved in the reaction. For reviews of the highly successful application of EFT($\not\pi$) in the PC sector, see, e.g., Refs. [5, 6, 7, 8].

At the low energies of interest, PV interactions between nucleons can be described by transitions between S- and P-wave states [9, 10]. Incorporating isospin, the leading-order (LO) PV EFT($\not\pi$) Lagrangian contains five independent operators [11, 12, 13]:

$$
\begin{aligned}
\mathcal{L}_{PV} = - \Big[& g^{(^3S_1-^1P_1)} d_t^{i\dagger} \left(N^T \sigma_2 \tau_2 \, i \overleftrightarrow{\nabla}_i N \right) \\
& + g^{(^1S_0-^3P_0)}_{(\Delta I=0)} d_s^{A\dagger} \left(N^T \sigma_2 \, \vec{\sigma} \cdot \tau_2 \tau_A \, i \overleftrightarrow{\nabla} N \right) \\
& + g^{(^1S_0-^3P_0)}_{(\Delta I=1)} \, \epsilon^{3AB} \, d_s^{A\dagger} \left(N^T \sigma_2 \, \vec{\sigma} \cdot \tau_2 \tau^B \overleftrightarrow{\nabla} N \right) \\
& + g^{(^1S_0-^3P_0)}_{(\Delta I=2)} \, \mathcal{I}^{AB} \, d_s^{A\dagger} \left(N^T \sigma_2 \, \vec{\sigma} \cdot \tau_2 \tau^B \, i \overleftrightarrow{\nabla} N \right) \\
& + g^{(^3S_1-^3P_1)} \, \epsilon^{ijk} \, d_t^{i\dagger} \left(N^T \sigma_2 \sigma^k \tau_2 \tau_3 \overleftrightarrow{\nabla}^j N \right) \Big] + \text{h.c.}
\end{aligned}
\tag{1}
$$

Here, d_t and d_s are auxiliary fields representing two-nucleon states in the $^3S_1$ and $^1S_0$ channel, respectively [14], $a\,\mathcal{O}\overleftrightarrow{\nabla}_i b = a\,\mathcal{O}D_i b - (D_i a)\mathcal{O}b$, with $\vec{D}$ a nucleon covariant derivative and $\mathcal{O}$ some spin-isospin-operator, and the isospin matrix $\mathcal{I} = \text{diag}(1, 1, -2)$. The $g^{(X-Y)}$ are the PV LECs encoding the short-distance physics. They cannot be determined within the EFT itself, but are related to standard model parameters though nonperturbative QCD calculations. Given the complexity of such calculations, another approach is to fit the LECs to data. However, this requires the existence of a sufficient number of experimental results and the calculation of the corresponding observables in the EFT($\not\pi$) framework.

3. Parity violation in two-nucleon systems

Because transition amplitudes can be determined analytically, two-nucleon systems present the simplest environments to determine PV observables theoretically, and a number of reactions have been considered. Using the scattering of a longitudinally polarized nucleon beam on an unpolarized nucleon target, a longitudinal asymmetry A_L can be defined as

$$A_L^{NN} = \frac{\sigma_+ - \sigma_-}{\sigma_+ + \sigma_-} \,, \tag{2}$$

where $\sigma_\pm$ represents the cross section for scattering of a nucleon beam with helicity ± 1. Results of a measurement of A_L^{pp} in proton-proton scattering at an energy low-enough for EFT($\not\pi$) to be applicable are reported in Ref. [15]. Neglecting Coulomb effects in the proton-proton case and adjusting conventions to be consistent with Eq (1), the LO EFT($\not\pi$) results are given by [16, 17, 12]

$$A_L^{nn} = -\sqrt{\frac{32M}{\pi}}\, p\left(g_{(\Delta I=0)}^{(^1S_0-^3P_0)} - g_{(\Delta I=1)}^{(^1S_0-^3P_0)} + g_{(\Delta I=2)}^{(^1S_0-^3P_0)}\right) , \tag{3}$$

$$A_L^{pp} = -\sqrt{\frac{32M}{\pi}}\, p\left(g_{(\Delta I=0)}^{(^1S_0-^3P_0)} + g_{(\Delta I=1)}^{(^1S_0-^3P_0)} + g_{(\Delta I=2)}^{(^1S_0-^3P_0)}\right) , \tag{4}$$

$$\begin{aligned} A_L^{np} = & -\sqrt{\frac{32M}{\pi}}\, p\frac{\frac{d\sigma^{^1S_0}}{d\Omega}}{\frac{d\sigma^{^1S_0}}{d\Omega} + 3\frac{d\sigma^{^3S_1}}{d\Omega}}\left(g_{(\Delta I=0)}^{(^1S_0-^3P_0)} - 2g_{(\Delta I=2)}^{(^1S_0-^3P_0)}\right) \\ & -\sqrt{\frac{32M}{\pi}}\, p\frac{\frac{d\sigma^{^3S_1}}{d\Omega}}{\frac{d\sigma^{^1S_0}}{d\Omega} + 3\frac{d\sigma^{^3S_1}}{d\Omega}}\left(g^{(^3S_1-^1P_1)} + 2g^{(^3S_1-^3P_1)}\right) , \end{aligned} \tag{5}$$

with p the momentum in the center-of-mass frame, M the nucleon mass, and $d\sigma/d\Omega$ the PC differential cross section in the corresponding channel. At the energy of the experiment of Ref. [15], Coulomb effects are shown to be of the order of approximately 3% and can therefore be safely neglected given the theoretical uncertainties of a LO calculation.

For a perpendicularly polarized neutron beam passing through an unpolarized target, PV interactions cause a rotation of the neutron spin. For a proton target with target density ρ, the rotation angle per unit length to next-to-leading (NLO) order is given by [18]

$$\frac{1}{\rho}\frac{d\phi_{\rm PV}^{np}}{dl} = 4\sqrt{2\pi}M\left(\frac{2g^{(^3S_1-^3P_1)} + g^{(^3S_1-^1P_1)}}{\gamma_t}\frac{Z_t+1}{2} + \frac{g_{(\Delta I=0)}^{(^1S_0-^3P_0)} - 2g_{(\Delta I=2)}^{(^1S_0-^3P_0)}}{\gamma_s}\frac{Z_s+1}{2}\right) , \tag{6}$$

where $Z_{t/s} = (1-\gamma_{t/s}r_{t/s})^{-1}$, and $\gamma_{t/s}$ ($r_{t/s}$) are the poles in the NN S-wave scattering amplitude (effective ranges) in the $^3\mathrm{S}_1/^1\mathrm{S}_0$ channel. An order-of-magnitude estimate of the size of the rotation angle per unit length for a typical hydrogen target is in the range 10^{-7} rad/m to 10^{-6} rad/m.

Deuteron photodisintegration and radiative neutron capture on protons provide several PV observables. Of particular experimental interest is the angular asymmetry A_γ in the capture of polarized neutrons ($\vec{n}p \to d\gamma$), which has been the goal of a measurement at the Spallation Neutron Source at Oak Ridge National Laboratory [19]. The LO EFT($\not\pi$) result is

$$A_\gamma = \frac{4}{3}\sqrt{\frac{2}{\pi}}\frac{M^{\frac{3}{2}}}{\kappa_1(1-\gamma_t a_s)}\, g^{(^3S_1-^3P_1)} \,, \tag{7}$$

where κ_1 is the nucleon isovector anomalous magnetic moment and a_s the $^1\mathrm{S}_0$ scattering length.

With ongoing developments in high-intensity photon sources, there is renewed interest in the measurement of the asymmetry A_γ^L in deuteron photodisintegration with polarized photons ($\vec{\gamma}d \to np$),

$$A_\gamma^L = \frac{\sigma_+ - \sigma_-}{\sigma_+ + \sigma_-}, \tag{8}$$

where $\sigma_\pm$ is the breakup cross section for a photon beam with helicity ± 1. For exactly reversed kinematics, this asymmetry is identical to the induced photon polarization P_γ in unpolarized neutron capture ($np \to d\vec{\gamma}$). It provides independent and complementary information to A_γ, as can be seen from the LO result for P_γ [13, 20],[1]

$$P_\gamma = -2\sqrt{\frac{2}{\pi}}\frac{M^{\frac{3}{2}}}{\kappa_1(1-\gamma_t a_s)}\left[\left(1-\frac{2}{3}\gamma_t a_s\right)g^{(^3S_1-^1P_1)} + \frac{\gamma_t a_s}{3}\left(g^{(^1S_0-^3P_0)}_{(\Delta I=0)} - 2g^{(^1S_0-^3P_0)}_{(\Delta I=2)}\right)\right]. \quad (9)$$

In particular, P_γ, or equivalently A^L_γ, is one of a very limited number of observables sensitive to the isotensor coupling $g^{(^1S_0-^3P_0)}_{(\Delta I=2)}$. Reference [21] determined the energy dependence of A^L_γ to NLO as well as a very simplistic figure of merit to determine at which energy to best perform the corresponding experiment. Because the PV LECs are currently not determined, Ref. [21] used different model estimates and found that the figure of merit is maximized for photon energies close to threshold, $2.259\text{MeV} < k < 2.264\text{MeV}$. However, these results only represent order-of-magnitude estimates because of the uncertainty in determining the LECs.

4. Parity violation in three-nucleon systems

In systems with three or more nucleons, three-nucleon (3N) forces have to be taken into account. The EFT power counting rules based on dimensional analysis indicate that 3N interactions are suppressed relative to NN interactions. However, in the PC sector of the theory a cutoff dependence is present in the LO solutions of the scattering amplitude for neutron-deuteron scattering in the spin doublet channel [22]. This unphysical dependence on the cutoff can be removed by the introduction of a 3N contact term at LO. The accompanying LEC can be related to a three-body observable, such as a nucleon-deuteron scattering length or a three-body binding energy. Given the difficulty in obtaining a sufficient number of experimental results to pin down the PV two-nucleon LECs, an analogous enhancement of a 3N interaction in the PV sector would present a significant obstacle in describing hadronic parity violation in few-nucleon systems.

In Ref. [23] the ultraviolet behavior of the PV neutron-deuteron scattering amplitude for S-P wave transitions is analyzed. It is shown that no divergent contribution appears at LO, while at NLO the spin-isospin structures of the allowed PV 3N operators are different from those of possibly divergent terms. Therefore, up to and including NLO, no PV 3N operators are required, and up to an accuracy of roughly 10% two-nucleon interactions should be sufficient to analyze parity violation in few-nucleon systems.

Considering a deuteron target, the spin rotation angle of a polarized neutron beam to NLO is EFT($\not\pi$) is given by [18] (also see Ref. [24] for a LO calculation)

$$\begin{aligned}\frac{1}{\rho}\frac{d\phi^{nd}_{\rm PV}}{dl} = \Big([16\pm1.6]g^{(^3S_1-^1P_1)} + [34\pm3.4]g^{(^3S_1-^3P_1)} \\ + [4.6\pm1.0](3g^{(^1S_0-^3P_0)}_{(\Delta I=0)} - 2g^{(^1S_0-^3P_0)}_{(\Delta I=1)})\Big)\frac{\text{rad}}{\text{MeV}^{\frac{1}{2}}}\,.\end{aligned} \quad (10)$$

Theoretical uncertainties are estimated using three different methods, and the reported errors are the most conservative of these three estimates. An order of magnitude estimate of the rotation angle per unit length is again in the range of 10^{-7} rad/m to 10^{-6} rad/m.

5. Parity violation in chiral EFT

The range of applicability of EFT($\not\pi$) is restricted to very low energies and to few-nucleon systems. To analyze processes involving higher energies and/or more complex nuclei, the

[1] Ref. [20] includes a formally higher-order contribution.

theoretical description has to include pions as dynamical degrees of freedom. Effective field theory methods can also be applied in this case. The LO PV pion-nucleon Lagrangian was constructed in Refs. [25, 26]. These results can be used to derive a PV nucleon-nucleon potential based on chiral symmetry [16, 27]. In addition to a one-pion-exchange contribution at LO, at NLO contact terms analogous to those in EFT($\not\pi$) and two-pion-exchange contributions appear. In addition to two-nucleon observables, chiral EFT has been used to determine the longitudinal asymmetry in the charge-exchange reaction $\vec{n}\,{}^3\mathrm{He} \to p^3\mathrm{H}$ [27, 28].

6. Parity-violating interactions in the large-N_c expansion

Given the challenges in obtaining experimental results to determine the PV LECs, additional theoretical constraints on the couplings can be very valuable. QCD in the limit of the number of colors N_c becoming large [29, 30] provides one method of arriving at such constraints. In order to apply the large-N_c analysis to two-nucleon interactions, the NN potential is defined by [31]

$$V(\mathbf{p}_-, \mathbf{p}_+) = \langle (\mathbf{p'}_1, C), (\mathbf{p'}_2, D) | \hat{H} | (\mathbf{p}_1, A), (\mathbf{p}_2, \beta, B) \rangle, \tag{11}$$

where A, B, C, D collectively denote nucleon spin and isospin quantum numbers. The momenta $\mathbf{p}_\pm$ are related to the momenta of the incoming and outgoing nucleons through $\mathbf{p}_\pm \equiv \mathbf{p}' \pm \mathbf{p}$, where $\mathbf{p}' = \mathbf{p'}_1 - \mathbf{p'}_2$ and $\mathbf{p} = \mathbf{p}_1 - \mathbf{p}_2$. The Hartree Hamiltonian is given by [30, 31]

$$\hat{H} = N_c \sum_n \sum_{s,t} v_{stn} \left(\frac{\hat{S}}{N_c}\right)^s \left(\frac{\hat{I}}{N_c}\right)^t \left(\frac{\hat{G}}{N_c}\right)^{n-s-t}, \tag{12}$$

where

$$\hat{S}^i = \hat{q}^\dagger \frac{\sigma^i}{2} \hat{q}, \quad \hat{I}^a = \hat{q}^\dagger \frac{\tau^a}{2} \hat{q}, \quad \hat{G}^{ia} = \hat{q}^\dagger \frac{\sigma^i \tau^a}{4} \hat{q}, \tag{13}$$

and $\hat{q}$ denotes the light-quark doublet. The matrix elements of these operators scale as [32, 33]

$$\langle N' | \hat{S} | N \rangle \sim \langle N' | \hat{I} | N \rangle \sim 1, \quad \langle N' | \hat{G} | N \rangle \sim \langle N' | \mathbb{1} | N \rangle \sim N_c. \tag{14}$$

in the large-N_c limit. The coefficients v_{stn} contain factors of momentum, and momenta scale as $\mathbf{p}_- \sim 1$ and $\mathbf{p}_+ \sim N_c^{-1}$ [32, 33, 31]. When applied to PV interactions, the large-N_c analysis leads to a hierarchy of terms [34, 35]. The LO terms are given by

$$V_{N_c}^{\not p} = N_c \left[U_{\not p}^1(\mathbf{p}_-^2) \mathbf{p}_- \cdot (\vec{\sigma}_1 \times \vec{\sigma}_2) \vec{\tau}_1 \cdot \vec{\tau}_2 + U_{\not p}^2(\mathbf{p}_-^2) \mathbf{p}_- \cdot (\vec{\sigma}_1 \times \vec{\sigma}_2) [\tau_1 \, \tau_2]_2^{zz} \right], \tag{15}$$

with $[\ldots]_2$ the symmetric and traceless rank-two tensor, and the coefficient functions $U_{\not p}^i(\mathbf{p}_-^2)$ scaling as $\mathcal{O}(1)$. There are four terms at NLO in the large-N_c expansion that are all isovector, while the six operators at next-to-next-to-leading order are isoscalar and isotensor like the LO terms.

These results have important implications when combined with the EFT($\not\pi$) approach [35]. While in the EFT power counting all five LECs in Eq. (1) are expected to be of the same size, the large-N_c analysis implies that only two of these terms are dominant in a combined EFT($\not\pi$) and large-N_c expansion. In addition, the large-N_c analysis implies a relation between the two isoscalar LECs that should hold up to corrections of order N_c^{-2}, or roughly 10%. The analysis of Ref. [35] shows that these results are not inconsistent with existing measurements.

The reduction in the number of dominant terms and the relation between the LECs in the large-N_c limit provide significant constraints on PV interactions. In particular the fact that the isotensor term is LO in the combined EFT and large-N_c analysis indicates the importance of finding corresponding experimental constraints and to perform a lattice QCD calculation of this term. While in principle the existing measurement of the longitudinal asymmetry in $\vec{p}p$ scattering at 221 MeV [36, 37] could provide information on the isotensor coupling, this energy is far outside the range of applicability of EFT($\not\pi$). At low energies, a measurement of the PV asymmetry in $\vec{\gamma}d \to np$ could put important experimental constraints on the isotensor component of the PV interaction.

7. Conclusions

Hadronic parity violation originates in an interplay of weak and nonperturbative strong interactions between quarks inside hadrons. It therefore provides a unique probe of our understanding of the standard model. While experimentally more accessible in complex nuclear systems, the recent focus has been on parity violation in few-nucleon systems, which can be more readily described in terms of two- and three-nucleon interactions. Effective field theories provide a model independent framework to systematically analyze and interpret experimental results and to provide theoretical error estimates. At LO in EFT($\not\pi$), which is applicable at the very low energies relevant to many of the recent experiments, parity violation is described in terms of five S-P transition operators. A variety of observables has been calculated within this framework. In addition, the large-N_c expansion of QCD provides important theoretical constraints on the PV interactions. In a combined EFT($\not\pi$) and large-N_c expansion, the number of LO terms reduces from five to two, and a relation between the two isoscalar LECs is predicted. These results provide strong motivation to determine the isotensor component of the PV interactions, both from a measurement of the longitudinal asymmetry in $\vec{\gamma}d \to np$ and from lattice QCD.

Acknowledgments

I thank H. W. Grießhammer, D. R. Phillips, R. P. Springer, and J. Vanasse for their collaboration and many interesting and stimulating discussions. This material is based upon work supported by the U.S. Department of Energy, Office of Science, Office of Nuclear Physics, under Award No. DE-SC0010300.

References

[1] Bowman C, Bowman J and Yuan V 1989 *Phys.Rev.* **C39** 1721–1724
[2] Haxton W C and Holstein B R 2013 *Prog.Part.Nucl.Phys.* **71** 185–203 (*Preprint* 1303.4132)
[3] Schindler M R and Springer R P 2013 *Prog.Part.Nucl.Phys.* **72** 1–43 (*Preprint* 1305.4190)
[4] Desplanques B, Donoghue J F and Holstein B R 1980 *Annals Phys.* **124** 449
[5] Beane S R, Bedaque P F, Haxton W C, Phillips D R and Savage M J 2000 *In *Shifman, M. (ed.): At the frontier of particle physics, vol. 1** 133–269. (*Preprint* nucl-th/0008064)
[6] Bedaque P F and van Kolck U 2002 *Ann. Rev. Nucl. Part. Sci.* **52** 339–396 (*Preprint* nucl-th/0203055)
[7] Platter L 2009 *Few Body Syst.* **46** 139–171 (*Preprint* 0904.2227)
[8] Vanasse J 2016 *Int. J. Mod. Phys.* **E25** 1641002 (*Preprint* 1609.03086)
[9] Danilov G 1965 *Phys.Lett.* **18** 40–41
[10] Danilov G 1971 *Phys.Lett.* **B35** 579–580
[11] Girlanda L 2008 *Phys. Rev.* **C77** 067001 (*Preprint* 0804.0772)
[12] Phillips D R, Schindler M R and Springer R P 2009 *Nucl. Phys.* **A822** 1–19 (*Preprint* 0812.2073)
[13] Schindler M R and Springer R P 2010 *Nucl. Phys.* **A846** 51–62 (*Preprint* 0907.5358)
[14] Kaplan D B 1997 *Nucl. Phys.* **B494** 471–484 (*Preprint* nucl-th/9610052)
[15] Eversheim P D, Schmitt W, Kuhn S E, Hinterberger F, von Rossen P *et al.* 1991 *Phys.Lett.* **B256** 11–14
[16] Zhu S L, Maekawa C M, Holstein B R, Ramsey-Musolf M J and van Kolck U 2005 *Nucl.Phys.* **A748** 435–498 (*Preprint* nucl-th/0407087)
[17] Holstein B R 2005 *Fizika.* **B14** 165–216 (*Preprint* nucl-th/0607038)
[18] Grießhammer H W, Schindler M R and Springer R P 2012 *Eur.Phys.J.* **A48** 7 (*Preprint* 1109.5667)
[19] Gericke M T, Alarcon R, Balascuta S, Barron-Palos L, Blessinger C *et al.* 2011 *Phys.Rev.* **C83** 015505
[20] Shin J, Ando S and Hyun C 2010 *Phys.Rev.* **C81** 055501 (*Preprint* 0907.3995)
[21] Vanasse J and Schindler M R 2014 *Phys. Rev.* **C90** 044001 (*Preprint* 1404.0658)

[22] Bedaque P F, Hammer H W and van Kolck U 2000 *Nucl. Phys.* **A676** 357–370 (*Preprint* nucl-th/9906032)
[23] Grießhammer H W and Schindler M R 2010 *Eur. Phys. J.* **A46** 73–83 (*Preprint* 1007.0734)
[24] Vanasse J 2012 *Phys.Rev.* **C86** 014001 (*Preprint* 1110.1039)
[25] Kaplan D B and Savage M J 1993 *Nucl. Phys.* **A556** 653–671 [Erratum-ibid. A **570**, 833 (1994)] [Erratum-ibid. A **580**, 679 (1994)]
[26] Savage M J and Springer R P 1998 *Nucl.Phys.* **A644** 235–244 erratum-ibid. A **657**, 457 (1999) (*Preprint* nucl-th/9807014)
[27] Viviani M, Baroni A, Girlanda L, Kievsky A, Marcucci L E and Schiavilla R 2014 *Phys. Rev.* **C89** 064004 (*Preprint* 1403.2267)
[28] Viviani M, Schiavilla R, Girlanda L, Kievsky A and Marcucci L 2010 *Phys.Rev.* **C82** 044001 (*Preprint* 1007.2052)
[29] 't Hooft G 1974 *Nucl. Phys.* **B72** 461
[30] Witten E 1979 *Nucl. Phys.* **B160** 57
[31] Kaplan D B and Manohar A V 1997 *Phys.Rev.* **C56** 76–83 (*Preprint* nucl-th/9612021)
[32] Dashen R F, Jenkins E E and Manohar A V 1995 *Phys. Rev.* **D51** 3697–3727 (*Preprint* hep-ph/9411234)
[33] Kaplan D B and Savage M J 1996 *Phys. Lett.* **B365** 244–251 (*Preprint* hep-ph/9509371)
[34] Phillips D R, Samart D and Schat C 2015 *Phys. Rev. Lett.* **114** 062301 (*Preprint* 1410.1157)
[35] Schindler M R, Springer R P and Vanasse J 2016 *Phys. Rev.* **C93** 025502 (*Preprint* 1510.07598)
[36] Berdoz A R *et al.* (TRIUMF E497) 2001 *Phys. Rev. Lett.* **87** 272301 (*Preprint* nucl-ex/0107014)
[37] Berdoz A *et al.* (TRIUMF E497 Collaboration) 2003 *Phys.Rev.* **C68** 034004 (*Preprint* nucl-ex/0211020)

Oxygen-15+α resonant elastic scattering to study cluster states in ^{19}Ne

D Torresi[1], C. Wheldon[1], Tz. Kokalova[1], S. Bailey[1], A. Boiano[2], C. Boiano[3], M. Cavallaro[4], S. Cherubini[4], A. Di Pietro[4], J.P. Fernandez Garcia[4], M. Fisichella[4], T.R. Glodariu[5], J. Grebosz[6], M. La Cognata[4], M. La Commara[2,7], M. Lattuada[4,8], M. Mazzocco[9,10], D. Mengoni[9,10], C. Parascandolo[2,7], D. Pierroutsakou[4], G. Pizzone[4,8], C. Signorini[10], C. Stefanini[9], L. Stroe[5], C. Spitaleri[4,8], E. Strano[9,10], and M. Zadro[11]

[1]School of Physics and Astronomy, University of Birmingham, Birmingham, B15 2TT, UK
[2]INFN-Sezione di Napoli, Via Cinthia, I-80126 Napoli, Italy
[3]INFN-Sezione di Milano, Via Celoria 16, I-20133 Milano, Italy
[4]INFN-Laboratori Nazionali del Sud, Via S. Sofia 62 Catania, Italy
[5]TNIPNE, 407 Atomistilor Street, 077125 Magurele, Romania
[6]Department of Nuclear Reactions, Institute for Nuclear Studies, ul. Hoza 69, 00-681 Warsaw, Poland
[7]Dipartimento di Fisica, Universitá di Napoli, Via Cinthia, I-80126 Napoli, Italy
[8]Dipartimento di Fisica, Universitá di Catania, Via A. Doria 64, I-95123 Catania, Italy
[9]INFN-Sezione di Padova, Via F. Marzolo 8, I-35131 Padova, Italy
[10]INFN-Laboratori Nazionali di Legnaro, Viale dellUniversitá 2, I-35020 Legnaro, Italy
[11]Rujder Bošković Institute, Bijenicka cesta 54, 10000 Zagreb, Croatia

E-mail: `domenico.torresi@lns.infn.it`

Abstract. The elastic scattering excitation function for ^{15}O on ^{4}He was measured - for the first time - by using the Inverse Kinematic Thick Target scattering method. The obtained spectrum was analysed in an R-matrix framework providing spectroscopic information for several states in the excitation energy of the nucleus ^{19}Ne from 4.8 to 9 MeV. Eight of them are newly observed states. The set-up and the experimental technique will be described in detail and the results will be shown and discussed.

1. Introduction

The first evidence of the presence of α cluster structures concerned nuclei which have even and equal numbers of protons and neutrons (so-called α-conjugate nuclei) like ^{8}Be, ^{12}C, ^{16}O. Probably one of the best examples of α cluster structure in α-conjugate nuclei is given by ^{20}Ne that is known to have a number of states associated with a well developed cluster structure comprising ^{4}He and ^{16}O closed-shell cores. It was found that these states can be grouped in several rotational bands [1, 2, 3]. Moreover the fact that these bands form inverse parity doublets offers further evidence of a mass asymmetric configuration: ^{4}He and ^{16}O. Also non-α-conjugate nuclei can manifest cluster structure, indeed, it is expected that – moving out of the valley of stability – light unstable nuclei may also exhibit cluster behaviour. Examples are ^{6}Li and ^{7}Li with pronounced α-d and α-t cluster structure in their ground state. The presence of the cluster configurations in the ground state of these nuclei is related to their very low binding energy, close to the cluster decay threshold.

The study of unstable systems presents many difficulties, due mainly, to the low intensities typical for radioactive ion beams. Therefore, few significant experimental studies have been performed so far. Studying unstable nuclei is difficult also because often the energy spectrum of these nuclei is poorly known. This difficulty can be partially overcome by comparing the energy spectrum of the nucleus under investigation with that of the associated mirror nucleus. The study of analogue states for isobars can indeed yield important spectroscopic information.
The present work shows results of the search for cluster structure in ^{19}Ne by means of resonant scattering using a radioactive beam ^{15}O on a ^{4}He gas target.

2. Experiment

2.1. Experimental technique

The Thick Target Inverse Kinematics scattering method (TTIK) with gas targets was firstly developed in Refs. [4] and [5] for the measurement of the elastic scattering excitation function. The main advantage of the technique is that it allows the elastic scattering cross-section to be measured over a wide energy range using a single beam energy. This characteristics make the technique particularly suitable for the use with Radioactive Ion Beams (RIBs) that usually have intensities orders of magnitude lower than stable beams. The technique consists of sending a beam on a gas target (typically a scattering chamber filled with helium) with such a thickness (pressure) so as to fully stop the beam before it reaches the detectors placed at and around zero degrees in the laboratory system. The gas acts at the same time as target and as degrader of the beam energy. The interaction between projectile and target nuclei can occur at different positions along the beam trajectory, and thus at different energies. Thanks to the different stopping powers of the gas for heavy projectiles and light recoils, the recoiling particles can reach a detector placed at zero degrees in the laboratory system, while the particles of the beam are stopped before reaching the detectors. From the measurement of the recoil energy, one can deduce the interaction position along the beam direction and the energy of the scattering event. This relationship is univocal, one is sure that the detected light recoils originate from elastic scattering. Thus, the Elastic Scattering Excitation Function, (ESEF) in a given energy range, can be measured by using a single beam energy. The extent of the energy range investigated depends, of course, on the initial beam energy.

2.2. Beam production

The experiment has been performed by using a ^{15}O RIB [6] produced by the in-flight facility EXOTIC [7, 8] at the Laboratori Nazionali di Legnaro (LNL, Italy). The facility uses primary heavy ion beams delivered by the XTU tandem Van de Graaff accelerator. The primary beam is incident on a cryogenic gas target, that can be filled with hydrogen, deuterium or ^{3}He gas, then direct reactions like (p,n), (d,n) (d,p) and $(^{3}\text{He},n)$ take place in the target and the reaction products are selected in flight and are finally sent to the secondary target.

For the production of the ^{15}O beam a primary $^{15}\text{N}^{5+}$ beam at energy of 80 MeV impinged on a H_2 cryogenic gas target at a pressure of 1 bar and a temperature of 93 K, corresponding to an equivalent target thickness of about 1.35 mg/cm^2 of H_2. The ^{15}O beam, produced by means of the two-body reaction $p(^{15}\text{N},^{15}\text{O})n$ ($Q_{val}=-3.54$ MeV), was selected and purified from the scattered primary ^{15}N beam and other contaminant nuclei produced in the interaction between the primary beam and the H_2 target by means of a set of optical elements and collimators. Two parallel-plate avalanche counters (PPACs) [9] and a set of silicon surface barrier detector were used for different purposes: the former to optimise the beam intensity and the latter for the beam diagnostic and on-line monitoring. The two PPACs were placed downstream prior to the

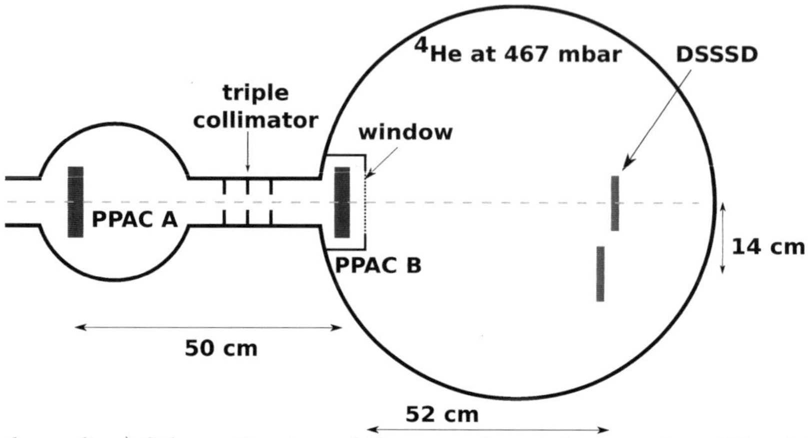

Figure 1. (color online) Schematic view of the experimental apparatus. The detection system consisted of one DSSSD and two PPACs required for beam counting and the time-of-flight measurement.

beam entering the scattering chamber for beam diagnostic purposes. The ^{15}O beam energy was 28.5 MeV and its intensity was in the range 1-2×10^4 particles/s with ≈99% purity.

2.3. Experimental apparatus

The experiment was performed using the scattering chamber of the EXOTIC beam line filled with ^{4}He gas at a pressure of around 467 mbar, enough to stop the ^{15}O beam at 28.5 MeV before reaching the silicon detector. The chamber was separated from the high vacuum beam line by a 2.2 μm thick Havar window. The pressure and the temperature of the helium gas were monitored by two capacitance manometers and a thermocouple device. It was verified that the temperature did not change appreciably during the whole experiment and that the two manometers were always giving consistent pressure values within their precision. The detection of the recoiling ^{4}He relied on a 6×6 cm^2 wide, 300 μm thick Double Sided Silicon Strip Detector (DSSSD) placed at zero degrees, that is part of a larger detection apparatus EXPADES [9]. Moreover two PPACs placed before the window were utilised for various purposes: first they were used for an on-line monitoring of the beam; second they provided the secondary ^{15}O beam profile by means of the event-by-event reconstruction of the trajectory; third for counting the beam particles and, finally, for timing purposes. The time-of-flight measurement between the entering of the ^{15}O particle and the following detection of a recoiling ^{4}He is used in the present experiment for the particle identification as will be shown in the next section. In Fig. 1 a sketch of the scattering chamber and the experimental apparatus is shown.

3. Data Analysis

The goal of the analysis is to deduce the elastic scattering excitation function in the centre of mass energy from the detected energy of the recoiling α particles. The first step in the analysis is to identify the recoiling α-particles and separate from other particles coming from reaction or as contaminants of the secondary beam. In Fig. 2 the plot of ToF versus detected energy is shown. The regions corresponding to α particles coming from elastic scattering and protons are easily identified. On top of the two dimensional spectrum a continuous red line is drawn that corresponds to a calculation of the ToF as a function of energy for elastically scattered α particles and a short dotted line corresponding to a calculation of the inelastic scattering to the excited states of the ^{15}O at 5.2 MeV. It can be clearly seen that the first excited state is scarcely

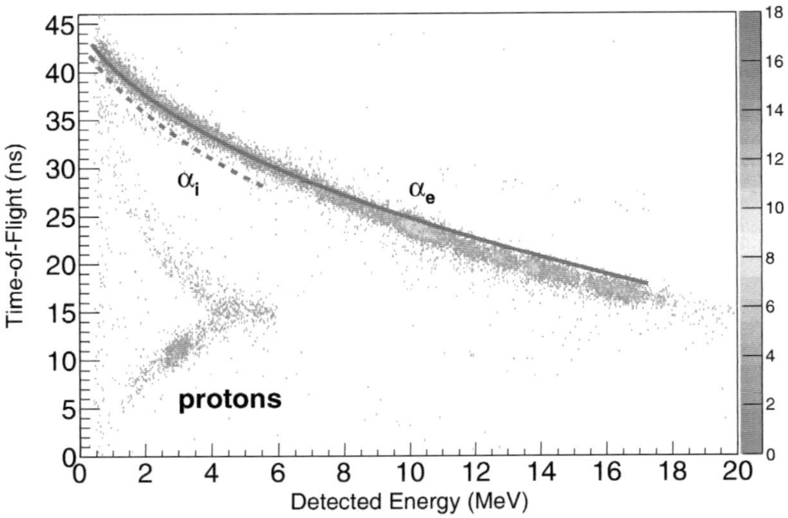

Figure 2. A plot of detected energy versus ToF. It is possible to distinguish α particles originating from an elastic scattering process and protons. The calculated distribution of α particles is indicated by the red continuous line. The red short-dashed line corresponds to α particles originating from inelastic scattering, demonstrating that events from this process have not been observed at a significant level. Adapted from Fig. 2 of Ref.[10].

populated or is not populated at all. Therefore, the ToF measurements in this case was used not only for the particle identification but also to make sure that inelastic scattering does not contaminate the elastic scattering spectrum.

Once the α particles corresponding to elastic scattering events are selected, through use of a graphical window 2D, the detected energy spectrum has to be transformed in ESEF. Indeed, the α particles are detected with an energy that is different from their energy immediately after the scattering, since they have to travel the distance between the scattering point and the detector that is filled with helium gas. The α particles lose an energy, ΔE, that is not known *a priori* since the scattering point is not known.

A calculation that associates each position at which the scattering events occurred to a corresponding energy of the α particles at the detector has been performed. This calculation requires as input, the geometry of the detection apparatus, the pressure and temperature of the gas and the stopping power of ^{4}He gas for ^{15}O and α particles. From this it follows that the knowledge of the stopping power is of fundamental importance to obtain a reliable elastic scattering excitation function. Indeed, the effect of using the wrong stopping powers does not simply shift the spectrum but deforms it, this has been demonstrated by Zadro *et al.* [11] where the elastic scattering of ^{9}Be on ^{4}He was measured and compared to the ESEF obtained using the stopping power of ^{4}He for ^{9}Be calculated by SRIM [12]. The importance of knowing the stopping power to high accuracy, for the current experiment is further illustrated by Fig. 3 where the same detected energy spectrum was transformed in ESEF using two different stopping power functions calculated using codes SRIM [12] and dedx [13]. The two different stopping power curves are shown in Fig. 4.

Due to the strong influence of the stopping power on the final ESEF, the stopping power for oxygen in helium gas was measured in a separate experiment [14], and the final ESEF, shown in Fig. 5 was obtained using this experimentally measured stopping power.

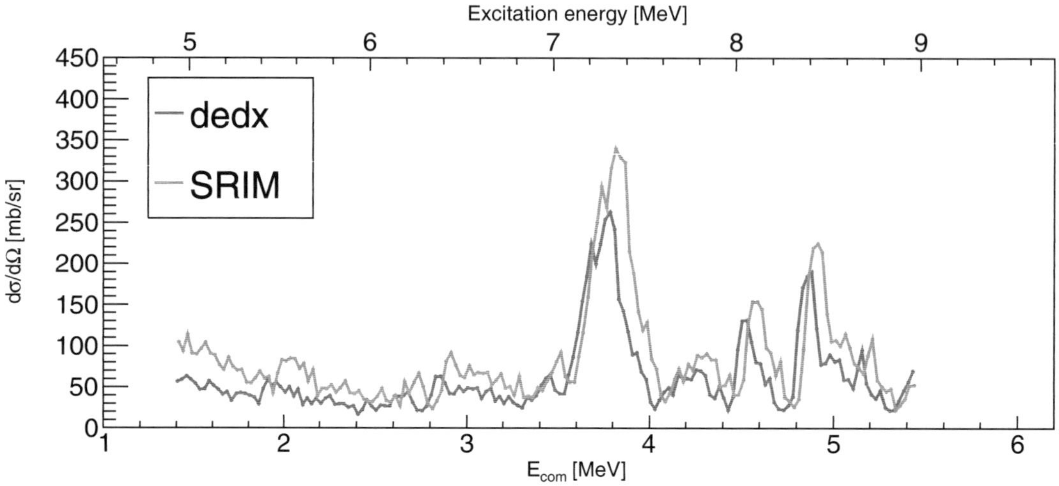

Figure 3. Elastic scattering excitation function deduced using two different stopping power functions calculated by the code SRIM (black curve) and the code *dedx* (green curve).

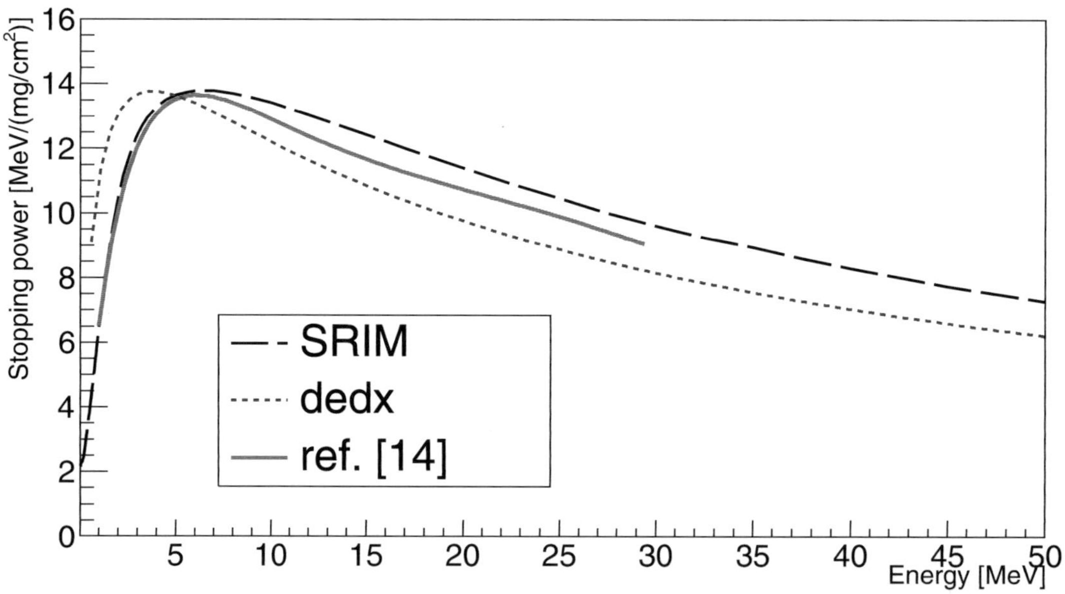

Figure 4. The stopping power curves for^{16}O in ^{4}He gas. The stopping power calculated by the code SRIM (black dashed line) and by the code *dedx* (dotted blu line) is compared with the experimentally measured one (red continuous line.

4. R-matrix calculation

In order to determine the main properties of the observed resonances: widths, partial decay branches, energies and spins, an R-matrix calculation was performed. A comprehensive description of the R-matrix theory can be found in Ref. [15]. A full R-Matrix fit was performed using Azure2 [16]. The spins and energies of states from the mirror nucleus, ^{19}F [17], were used as a starting point for the fit; a powerful technique, with a good agreement between the energy levels in each nucleus. A two channel calculation was performed to fit the data: ^{15}O+α and ^{18}F+p. Levels in the 5.3-8.9 MeV excitation-energy region have been investigated, with a number of newly observed states in this work. Furthermore, partial widths have been extracted for the observed levels, for many of them it is the first time measure-width data has been available.

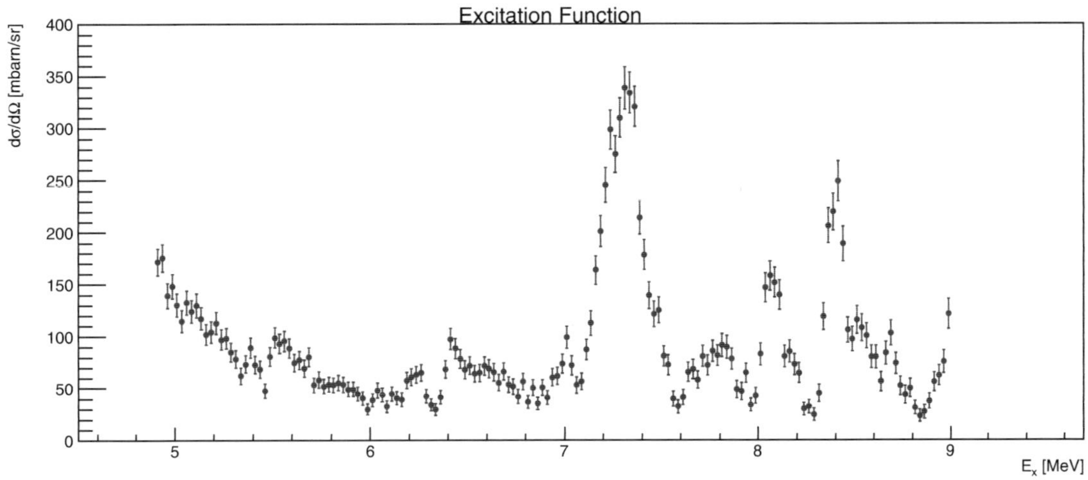

Figure 5. Elastic scattering excitation function deduced using the experimentally measured stopping power.

The results of R-matrix calculation will be discussed in details in a Ref. [10].

5. Conclusion

We have reported the measurements of the ^{15}O-α elastic scattering excitation function by using the TTIK method. The technique included a time-of-flight measurement that allows the discrimination of elastic scattering from other reaction mechanisms. The stopping power of the helium gas for the oxygen isotope was measured to avoid systematic errors that the use of a calculated stopping power can introduce in the deduction of the centre-of-mass energy from the detected energy. The excitation function was measured for excitation energies in the range of 5.3 to 8.9 MeV and it was analysed using an R-matrix approach. The R-matrix calculation, that reproduces rather well the experimental data, includes several new resonances with evidence for pronounced cluster structure [10].

Acknowledgements

The present work has received funding from the European Unions Horizon 2020 research and innovation programme under the Marie Sklodowska-Curie grant agreement No 659744. The UK STFC is acknowledged for providing funding under grant ST/L005751/1. M.Z. was partially supported by the Croatian Science Foundation under Project No. 7194.

References

[1] T. Tomoda and A. Arima, Nucl. Phys. A **303**, 217 (1978).
[2] S.R. Riedhauser *et al.*, Phys. Rev. C **29** (1984) 1961.
[3] H.T. Richards, Phys. Rev. C **29**, 276 (1984).
[4] K.P. Artemov *et al.*, Sov. J. Nucl. Phys. **52**, 634 (1990).
[5] K. Kallman *et al.*, Nucl. Instr. And Meth. in Phys. Res., A **338**, 413 (1994).
[6] D. Torresi, C. Wheldon *et al.*, LNL Annual Report (2014) p19, see *http://www.lnl.infn.it/~annrep/read_ar/2014/ contributions/pdfs/019_A_61_A056.pdf.*
[7] V.Z. Maidikov *et al.*, Nucl. Phys. A **746**, 389c (2004).
[8] M. Mazzocco *et al.*, Nucl. Instr. and Meth. in Phys. Res. B **317**, 223 (2013).
[9] D. Pierroutsakou *et al.*, Nucl. Instr. and Meth. in Phys. Res. A **834**, 46 (2016).
[10] D. Torresi *et al.*, Phys. Lett. B. submitted.
[11] M. Zadro *et al.*, Nucl. Instr. and Meth. in Phys. Res. B **259**, 836 (2007).
[12] J.F. Ziegler, Nucl. Instr. and Meth. in Phys. Res. B **219-220**, 1027 (2004).

[13] Computer Program DEDX (University of Birmingham, UK) unpublished. Based on computer program SPAR, T.W. Armstrong, K.C. Chandler, ORNL-4869 (Oak Ridge National Laboratory, US) 1973 and T.W. Armstrong, K.C. Chandler, Nuclear Instruments and Methods **113**, 313 (1973).
[14] D. Torresi *et al.*, Nucl. Instr. and Meth. in Phys. Res. B **389-390**, 1 (2016).
[15] A,M. Lane and R.G. Thomas, Reviews of Modern Physics **30**, 257 (1958).
[16] R. Azuma *et al.*, Phys. Rev. C **81**, 045805 (2010).
[17] D.R. Tilley, H.R. Weller, C.M. Cheves and R.M. Chasteler, Nucl. Phys A **595**, 1 (1995).

IOP Conf. Series: Journal of Physics: Conf. Series **876** (2017) 012022 doi:10.1088/1742-6596/876/1/012022

TDA and RPA pseudoscalar and vector solutions for the low energy regime of a motivated QCD Hamiltonian.

T. Yépez-Martínez[1], D. A. Amor Quiroz[2], P. O. Hess[2] and O. Civitarese[1]

[1] Departamento de Física, Universidad Nacional de La Plata, C.C.67 (1900), La Plata, Argentina

[2] Instituto de Ciencias Nucleares, UNAM, Circuito Exterior, C.U., A.P. 70-543, 04510 México, D.F., México.

E-mail: yepez@fisica.unlp.edu.ar, arturo.amor@nucleares.unam.mx, hess@nucleares.unam.mx, osvaldo.civitarese@fisica.unlp.edu.ar

Abstract. We present the low energy meson spectrum of a Coulomb gauge QCD motivated Hamiltonian for light and strange quarks. We have used the harmonic oscillator as a trial basis and performed a pre-diagonalization of the kinetic energy term in order to get an effective basis where quark and anti-quark degrees of freedom are defined. For the relevant interactions between quarks and anti-quarks, we have implemented a confining interaction between color sources, in order to account in an effective way for the gluonic degrees of freedom. The low energy meson spectrum is obtained from the implementation of the TDA and RPA many-body-methods. The physical states have been described as TDA and RPA collective states with a relatively good agreement. Particularly, the particle-hole correlations of the RPA ground state improve the RPA pion-like state (159.7 MeV) close to its physical value while the TDA one remains at a higher energy (269.2 MeV).

1. Introduction

The Quantum Chromodynamics (QCD) is the fundamental theory of strong interaction, the basic ingredients of this theory are the elementary particles, quarks and gluons. The QCD describes the propagation and interactions of these particles. The quarks and gluons are color charged particles, where due to the non-abelian nature of the theory, the gluons can interact with themselves. The latter is a very important characteristic since several interactions and phenomena are associated to this fact, *e.g.*, the confinement of quarks and the chiral symmetry breaking. The low energy regime of QCD is highly non-perturbative since the interaction coupling is large. However, the Lattice Gauge Theory (LGT) [1–4], the Dyson-Schwinger equation (DS-eq) [5–7] and the effective models [8–11] have been able to get some insights about this regime. Particularly, the determination of the spectrum of QCD at low energy has been a task for many years. In dealing with this problem, LGT has shown to be a very powerful tool. Unfortunately, it has several disadvantages like huge computational efforts, problems with the identification of the states with a given spin and the calculation of excited states. Recently, LGT has been able to determine the spectrum of light and more heavy [3,4] quarks in relatively good agreement with the available experimental data.

From the side of the phenomenological approaches, to work with an effective interaction has shown to be also a good tool in order to describe the spectrum at low energy. They can be questionable mainly because the origin of the interactions and how they are related to real QCD. The Coulomb gauge QCD formalism derived in [12], underlying the relevant interactions of QCD which are dominated by the instantaneous Coulomb potential (ICP) acting between color charges. It is known that a confinement interaction able to reproduce LGT potential results [13], can be obtained in this formalism by the implementation of self-consistent treatments like DS-eq in the Yang-Mills sector and under certain approximations of the vacuum functional of QCD [14–16]. Following that direction, we have focused on the quark sector of the Coulomb gauge QCD Hamiltonian and accounts for the effects of the gluon dynamics with an effective interaction. In the past, we have analysed, step by step, the implementation of a parametrized interaction with analytic and semi-analytic approaches as well as pairing methods like BCS [17, 18]. More recently, we have implemented the Tamm-Dancoff-Approximation (TDA) [19, 20] and the Random-Phase-Approximation (RPA) [21, 22] many-body methods, in order to describe meson states as collective states.

In this work, we have considered a more realistic interaction of the type Coulomb $(-a/|\mathbf{x}-\mathbf{y}|)$ plus linear $(b|\mathbf{x}-\mathbf{y}|)$ potential and implement both the TDA and RPA methods [23]. We have restricted the calculations to well defined sub-spaces of spin, parity (J^P), flavor-hypercharge and isospin (Y, T) and only color-singlet $(0,0)_1$ or physical states. Within these sub-spaces, we have analyzed two sets of parameters, where the quark masses and interaction parameter were adjusted in order to reproduce certain features of the low lying meson spectrum. Particularly, the set of parameters leading to a RPA-pion-like state close to the experimental value is used for comparison with other similar approaches [24] and with experimental data [25] .

2. Motivated QCD Hamiltonian and basic ingredients.

We consider a motivated QCD Hamiltonian which is built by the kinetic and mass terms together with a confining interaction between quark color charge densities,

$$H_{eff} = \int \left\{ \psi^\dagger(\mathbf{x})(-i\alpha\cdot\nabla + \beta m_0)\psi(\mathbf{x})\right\} d\mathbf{x} - \frac{1}{2}\int \rho_a(\mathbf{x}) V(|\mathbf{x}-\mathbf{y}|)\rho^a(\mathbf{y}) d\mathbf{x}d\mathbf{y}, \quad (1)$$

where $\rho^a(\mathbf{x}) = \rho^a_q(\mathbf{x}) = \psi^\dagger(\mathbf{x})T^a\psi(\mathbf{x})$ and

$$\begin{aligned} \psi^\dagger(\mathbf{x}) &= \sum_{\tau N l m_l \sigma c f} R^*_{Nl}(x) Y^*_{lm_l}(\widehat{\mathbf{x}})\chi^\dagger_\sigma \mathbf{b}^\dagger_{\tau,Nlm_l,\sigma c f} \\ &= \sum_{Njm_jlm_l,\sigma c f} R^*_{Nl}(x)\langle lm_l, \frac{1}{2}\sigma | jm_j\rangle Y_{lm_l}(\hat{\mathbf{x}})\chi^\dagger_\sigma \left(b^\dagger_{-\frac{1}{2}(Nl)jm_j,cf} + b^\dagger_{\frac{1}{2}(Nl)jm_j,cf}\right) . \end{aligned} \quad (2)$$

In the last step, we have used the coupled representation of angular momentum and intrinsic spin-$\frac{1}{2}$ to total spin (j). The index τ identifies the creation(annihilation) of particles with positive $\tau = \frac{1}{2}$ and negative energy $\tau = -\frac{1}{2}$. The color and flavor intrinsic degrees of freedom are denoted by c and f, where f is a short hand notation for flavor hypercharge, isospin and third component of isospin $\{Y, T, T_z\}$. Since we are going to consider that the chiral and flavor symmetries are broken ($m_0 \neq 0$ and $m_0 \to m_0^T$), the mass of the light up and down $(T = \frac{1}{2})$ quarks are different than the strange $(T = 0)$ quark $(m_{u,d} \neq m_s)$.

Because of confinement, the domain of fields in Eq. (2) is expected to be restricted to a finite volume in space. Therefore, the eigenfunctions of the three dimensional harmonic oscillator are chosen as the single-particle orbitals.

$$R_{Nl}(x) = N_{Nl}\exp(-\frac{B_0 x^2}{2}) r^l L^{l+\frac{1}{2}}_{\frac{N-l}{2}}(B_0 x^2) \ , \quad (3)$$

where N refers to the number of oscillation-quanta of the orbital level and l the angular momentum. The particles are restricted to a finite volume of the size of the oscillator length $(\frac{1}{\sqrt{B_0}})$.

The use of the harmonic oscillator basis for the orbital part seems to restrict the validity of the approximations to a non-relativistic theory, but this is only true when individual levels are considered to describe quark states. When the complete basis is used, any relativistic state can be expanded into the non-relativistic basis. This requires to expand the relativistic states into, in general, an infinite sum. When sufficient basis states are included, this sum can be limited in such a way that adding new terms would not modify the results. The latter is related with a non-trivial renormalization procedure [22]. Here, we explore the low energy meson spectrum in terms of the basic ingredients.

2.1. *The Hamiltonian matrix elements.*

Using the fermion field quatization given in Eq. (2), the Hamiltonian of Eq. (1) is expressed as

$$\begin{aligned} H_{eff} &= \sum_{jT} \sum_{\tau_i N_i l_i} K^{j,T}_{\tau_1(N_1 l_1),\tau_2(N_2 l_2)} \left(\boldsymbol{b}^{\dagger}_{\tau_1(N_1,l_1)jT} \cdot \boldsymbol{b}^{\tau_2(N_2,l_2)jT} \right) \\ &+ \sum_{\{N_i l_i j_i;L\}} V^{L}_{\{N_i l_i j_i\}} \left[\left[b^{\dagger}_{(N_1 l_1)j_1 T_1} \otimes b_{(N_2 l_2)j_2 T_2} \right]^{\gamma} \otimes \left[b^{\dagger}_{(N_3 l_3)j_3 T_3} \otimes b_{(N_1 l_1)j_4 T_4} \right]^{\gamma} \right]^{\hat{0}} \end{aligned} \tag{4}$$

where $\gamma = 0, L, (1,1)_1, 0, 0$ indicates the intermediate couplings of pseudo-spin, spin ($L = 0, 1$), $(1,1)_1$-color and flavor hypercharge and isospin quantum numbers. The scalar structure of the interaction is denoted by $\hat{0} = 0, (0,0)_1, 0$, representing the total spin, color and flavor hypercharge and isospin couplings.

The matrix elements of the kinetic-energy are given by

$$K^{j,T}_{\tau_1(N_1 l_1),\tau_2(N_2 l_2)} = \begin{cases} k^{j}_{N_1 N_2} & \text{if } l_1 = j+\frac{1}{2}\,,\ l_2 = j-\frac{1}{2}\,,\ \tau_1 \neq \tau_2 \\ k^{j}_{N_2 N_1} & \text{if } l_1 = j-\frac{1}{2}\,,\ l_2 = j+\frac{1}{2}\,,\ \tau_1 \neq \tau_2 \\ m^{T}_{0} & \text{if } (N_1 l_1) = (N_2 l_2)\,,\ \tau_1 = \tau_2 = +\frac{1}{2} \\ -m^{T}_{0} & \text{if } (N_1 l_1) = (N_2 l_2)\,,\ \tau_1 = \tau_2 = -\frac{1}{2} \\ 0 & \text{in all other cases} \end{cases} \tag{5}$$

with $m^T_0 = m_{u,d}\delta_{T,\frac{1}{2}} + m_s \delta_{T,0}$ and

$$k^{j}_{N_1 N_2} = \sqrt{B_0}\sqrt{\frac{N_1 - j + \frac{3}{2}}{2}}\delta_{N_2,N_1+1} + \sqrt{B_0}\sqrt{\frac{N_1 + j + \frac{3}{2}}{2}}\delta_{N_2,N_1-1} \tag{6}$$

while the potential matrix elements are given by

$$\begin{aligned} & V^{L}_{\{N_i l_i j_i\}} \\ &= \sum_{J N'_r N_r l_r N_R L_R} 3\sqrt{8}\sqrt{(2j_1+1)(2j_2+1)(2j_3+1)(2j_4+1)}\sqrt{2L+1}(2J+1) \\ &\times (-1)^{L+j_2+j_4-J+1} \left\{ \begin{array}{ccc} l_1 & L & l_2 \\ j_2 & \frac{1}{2} & j_1 \end{array} \right\} \left\{ \begin{array}{ccc} l_3 & L & l_4 \\ j_4 & \frac{1}{2} & j_3 \end{array} \right\} \left\{ \begin{array}{ccc} l_2 & J & l_4 \\ l_3 & L & l_1 \end{array} \right\} \\ &\times (N'_r l_r, N_R L_R, J \mid N_1 l_1, N_3 l_3, J)(N_r l_r, N_R L_R, J \mid N_2 l_2, N_4 l_4, J) \int d^3 r \Psi^{*}_{N'_r l'_r l_r}(\mathbf{r}) V(\sqrt{2}r) \Psi_{N_r l_r l_r}(\mathbf{r}) \quad . \end{aligned} \tag{7}$$

The potential $V(\sqrt{2}r)$ represents the confining potential in term of the relative coordinate $\mathbf{r} = \frac{1}{\sqrt{2}}(\mathbf{x}-\mathbf{y})$. In Eq. (7), we have used the Moshinsky brackets (round-brackets) [26, 27] and integrated the center of mass coordinate $\mathbf{R} = \frac{1}{\sqrt{2}}(\mathbf{x}+\mathbf{y})$. The detailed derivation can be found in [22].

3. The pre-diagonalization of the kinetic term and the effective basis.

The diagonalization of the kinetic term (Eq. (5)) is achieved by a general procedure *i.e.*, we implement a unitary basis transformation

$$\boldsymbol{b}^{\dagger}_{\tau(N,l)jm_jcTT_z} = \sum_{\lambda\pi k}\left(\alpha^{jT}_{\tau(N,l);\lambda\pi k}\right)^{*}\boldsymbol{b}^{\dagger}_{\lambda\pi kjm_jcTT_z} \tag{8}$$

where the kinetic term has a diagonal representation

$$\sum_{\tau_i,N_i,l_i}\left(\alpha^{j,T}_{\tau_1,(N_1,l_1),\lambda_1\pi_1k_1}\right)^{*}K^{j,T}_{\tau_1(N_1l_1),\tau_2(N_2l_2)}\alpha^{j,T}_{\tau_2,(N_2,l_2),\lambda_2\pi_2k_2} = \varepsilon_{\lambda_1,\pi_1,k_1,j,T}\;\delta_{\lambda_1,\lambda_2}\delta_{\pi_1,\pi_2}\delta_{k_1,k_2}\,. \tag{9}$$

The corresponding eigenvalues $(\varepsilon_{\lambda_1,\pi_1,k_1,j,T})$ and eigenvectors $\left(\alpha^{jT}_{\tau(N,l);\lambda\pi k}\right)$ allow us to rewrite the Hamiltonian in terms of creation(annihilation) effective $b^{\dagger}_{\Gamma_i}(b_{\Gamma_i})$ quarks $(\lambda_i = 1/2)$ and $d^{\dagger}_{\Gamma_i}(d_{\Gamma_i})$ anti-quarks $(\lambda_i = -1/2)$ operators:

$$\begin{aligned} H_{eff} &= \sum_{\pi,k,j,Y,T}\varepsilon_{\pi,k,j,Y,T}\Big(\boldsymbol{n}_{\pi,k,j,Y,T}+\bar{\boldsymbol{n}}_{\pi,k,j,Y,T}\Big) \\ &- \frac{1}{2}\sum_{L}\sum_{\{\gamma_i\}}V^{L}_{\{\gamma_i\}}\Big\{\Big[\Big(\big[\boldsymbol{b}^{\dagger}_{\gamma_1}\otimes\boldsymbol{b}_{\gamma_2}\big]^{\gamma}-\big[\boldsymbol{d}_{\gamma_1}\otimes\boldsymbol{d}^{\dagger}_{\gamma_2}\big]^{\gamma}\Big)\otimes\Big(\big[\boldsymbol{b}^{\dagger}_{\gamma_3}\otimes\boldsymbol{b}_{\gamma_4}\big]^{\gamma}-\big[\boldsymbol{d}_{\gamma_3}\otimes\boldsymbol{d}^{\dagger}_{\gamma_4}\big]^{\gamma}\Big)\Big]^{\hat{0}} \\ &+ \Big[\Big(\big[\boldsymbol{d}_{\gamma_1}\otimes\boldsymbol{b}_{\gamma_2}\big]^{\gamma}-\big[\boldsymbol{b}^{\dagger}_{\gamma_1}\otimes\boldsymbol{d}^{\dagger}_{\gamma_2}\big]^{\gamma}\Big)\otimes\Big(\big[\boldsymbol{d}_{\gamma_3}\otimes\boldsymbol{b}_{\gamma_4}\big]^{\gamma}-\big[\boldsymbol{b}^{\dagger}_{\gamma_3}\otimes\boldsymbol{d}^{\dagger}_{\gamma_4}\big]^{\gamma}\Big)\Big]^{\hat{0}}\Big\} \end{aligned} \tag{10}$$

where $\boldsymbol{n}_{\pi,k,j,Y,T}$ and $\bar{\boldsymbol{n}}_{\pi,k,j,Y,T}$ are the quark and anti-quark number operators, $\gamma_i = \pi_ik_ij_iY_iT_i$ and $V^{L}_{\{\gamma_i\}} = V^{L}_{\{\lambda_i,\pi_i,k_i,j_i,Y_i,T_i\}}$ with

$$\begin{aligned} V^{L}_{\{\lambda_i,\pi_i,k_i,j_i,Y_i,T_i\}} &= \sum_{\tau_i,N_i,l_i}V^{L}_{\{N_il_ij_i\}}\,\alpha^{j_1,T_1}_{\tau_1(N_1l_1),\lambda_1,\pi_1,k_1}\alpha^{j_2,T_2}_{\tau_2(N_2l_2),\lambda_2,\pi_2,k_2}\alpha^{j_3,T_3}_{\tau_3(N_3l_3),\lambda_3,\pi_3,k_3}\alpha^{j_4,T_4}_{\tau_4(N_4l_4),\lambda_4,\pi_4,k_4} \\ &\times\;\delta_{\tau_1\tau_2}\delta_{\tau_3\tau_4}\,(-1)^{\frac{1}{3}+\frac{Y_1}{2}+T_1}\frac{\sqrt{2T_1+1}}{\sqrt{3}}\delta_{T_2T_1}\delta_{Y_2-Y_1}\,(-1)^{\frac{1}{3}+\frac{Y_3}{2}+T_3}\frac{\sqrt{2T_3+1}}{\sqrt{3}}\delta_{T_4T_3}\delta_{Y_4-Y_3} \end{aligned} \tag{11}$$

The new extra factor $(-1)^{\frac{1}{3}+\frac{Y_1}{2}+T_1}\frac{\sqrt{2T_1+1}}{\sqrt{3}}\delta_{T_2T_1}\delta_{Y_2-Y_1}$ appears as a consecuence of the flavor symmetry breaking and it comes from extracting the flavor isoscalar of SU(3) [28]. In Eq. (10), all the information of the chiral and flavor symmetry breaking is carried by the coeficients $\alpha^{j,T}_{\tau(Nl),\lambda,\pi,k}$ of the unitary transformation Eq. (8).

4. The TDA and RPA pseudoscalar and vector solutions.

We implement the TDA and RPA methods [19–23], to define collective states as a linear combination of the pairs, built by quark $(b^{\dagger}_{\gamma_i})$ and anti-quark $(d^{\dagger}_{\gamma_i})$. The creation RPA collective operator is given by

$$\hat{\Gamma}^{\dagger}_{n;\Gamma\mu} = \sum_{\gamma_a,\gamma_b} \left\{ X^{n}_{\gamma_a,\gamma_b;\Gamma} [b^{\dagger}_{\gamma_a} d^{\dagger}_{\gamma_b}]^{\Gamma}_{\mu} - Y^{n}_{\gamma_a,\gamma_b;\Gamma} (-1)^{\phi_{\Gamma\mu}} [d^{\gamma_b} b^{\gamma_a}]^{\Gamma}_{\bar{\mu}} \right\} \tag{12}$$

where we have used the short hand notation $\Gamma = J^P, (0,0)_1, Y, T$ to denote the spin-parity, color and flavor hypercharge and isospin of the collective operator and $\mu = M_J, 0, T_z$. The index n labels the new collective state and it runs from zero to the number of possibilities to form pairs with quantum numbers $\Gamma\mu$. The backward amplitudes $Y^{n}_{\gamma_a,\gamma_b;\Gamma} \neq 0$ imply the possibility of annihilate pairs from the ground state and in that sense the RPA method accounts for particle-hole correlations on both: the ground state and the excited states. The n-th RPA excited state is obtained by $|n;\Gamma\mu\rangle = \hat{\Gamma}^{\dagger}_{n;\Gamma\mu}|RPA\rangle$ and the RPA-ground-state ($|RPA\rangle$) is constructed in such a way that the condition $\hat{\Gamma}^{n;\Gamma\mu}|RPA\rangle = 0$ is fulfilled.

The TDA collective operator is obtained from the RPA one by setting to zero the backward amplitudes $Y^{n}_{\gamma_a,\gamma_b;\Gamma} = 0$. Thus, the n-th TDA excited state is obtained by $|n;\Gamma\mu\rangle = \hat{\Gamma}^{\dagger}_{n;\Gamma\mu}|\tilde{0}\rangle$, where the TDA-ground-state ($|\tilde{0}\rangle$) is annihilated by the quark and anti-quark operators $b_{\gamma_i}|\tilde{0}\rangle = d_{\gamma_i}|\tilde{0}\rangle = 0$.

In Table 1, we show how the subspaces defined by quantum numbers of the collective states $J^P, (0,0)_1, Y, T$ are related with the physical pseudoscalar and vector mesons. In particular, the subspaces with $Y, T = 0, 0$, do not show any flavor mixing between pure $|q\bar{q}\rangle$ ($q = up, down$ quarks) and $|s\bar{s}\rangle$ states, the latter is due to the color confining interaction. Therefore, in order to make a closer description of physical mesons like the pseudoscalars η and η' or the vector ω and ϕ mesons, a flavor mixing (FM) should be implemented.

Meson	$\{J^P, (0,0)_1, Y, T\}$
	Pseudoscalar
π	$\{0^-, (0,0)_1, 0, 1\}$
K	$\{0^-, (0,0)_1, \pm 1, \frac{1}{2}\}$
η, η'	$\{0^-, (0,0)_1, 0, 0\}$+FM
	Vector
ρ	$\{1^-, (0,0)_1, 0, 1\}$
K^*	$\{1^-, (0,0)_1, \pm 1, \frac{1}{2}\}$
ω, ϕ	$\{1^-, (0,0)_1, 0, 0\}$+FM

Table 1. Identification of the collective-state-subspaces with physical mesons.

In these sample calculations, we have restricted to total single particle spin $j = \frac{1}{2}$, and take the maximal number of oscillation quanta $N_{max} = 3$. One feature we noted for this low energy calculation is that the $-\frac{a}{|\mathbf{x}-\mathbf{y}|}$ interaction does not have a big influence on the TDA and RPA results, which is due to the low energy regime where long-range ($b|\mathbf{x}-\mathbf{y}|$) effects are dominant. For higher maximal number of oscillation quanta N_{max}, the contribution of the interaction $-\frac{a}{|\mathbf{x}-\mathbf{y}|}$ becomes more important, as expected. However, when the cut-off, N_{max}, is increased, the calculations must be renormalized in order that the observables remain unchanged. The details about the renormalization of the quark masses $m_{u,d}, m_s$ and of the coupling constants a and b as a function of N_{max} are discussed in [22].

IOP Conf. Series: Journal of Physics: Conf. Series **876** (2017) 012022 doi:10.1088/1742-6596/876/1/012022

Set	$m_{u,d}[GeV]$	$m_s[GeV]$	$b[GeV^2]$
1	0.008	0.31	0.25
2	0.008	0.31	0.40

Table 2. Sets of parameters used for the TDA and RPA calculations. The a parameter is set to zero and the oscillator length $1/\sqrt{B_0}$ is fixed at 0.7 fm.

In Table 2, we show the parameters used in these sample calculations. The first set leads to a relatively heavy pion-like state while the second set leads to a pion-like state closer in energy to the experimental value. The selection of these two sets is motivated from the LGT, where one can find that the pion reported is at about 400 MeV [4] or at about 150 MeV [3]. In both case we explore how the TDA and RPA pseudoscalar and vector spectrum at about 1 GeV looks like. The results are presented in Table 3.

State	Exp.	Set-1(TDA)	Set-1(RPA)	Set-2(TDA)	Set-2(RPA)
π	139	392.8	369.3	269.2	159.7
K	495	580.1	580.1	490.5	490.5
η	547	467.0	485.2	402.1	402.3
η'	957	719.4	688.0	646.2	586.2
ρ	770	643.0	639.5	710.8	698.9
ω	782	695.9	691.3	784.5	782.6
K^*	892	852.9	852.9	966.1	966.1
ϕ	1020	1020.4	1020.4	1144.5	1135.5

Table 3. Comparison of pseudoscalar and vector mesons with the TDA and RPA results obtained using Sets 1 and 2 of parameters.

The two set of parameters of Table 1 are able to reproduce relatively well the meson spectrum at low energy. In the case of Set-1, the pion state is higher in energy compared to the experimental value, but the rest of the collective pseudoscalar and vector mesons are in good agreement with data. In the case of Set-2, the mass of the RPA pion-like state is closer to the experimental value, but the pseudoscalar states associated with $Y, T = 0, 0$ are lowered in energy. The latter is an indicator that in the present approach certain interactions are missing, *e.g.*, related with dynamical gluons. However, the present results show that observed energy differences in spin and isospin are satisfactorily reproduced, to some extent.

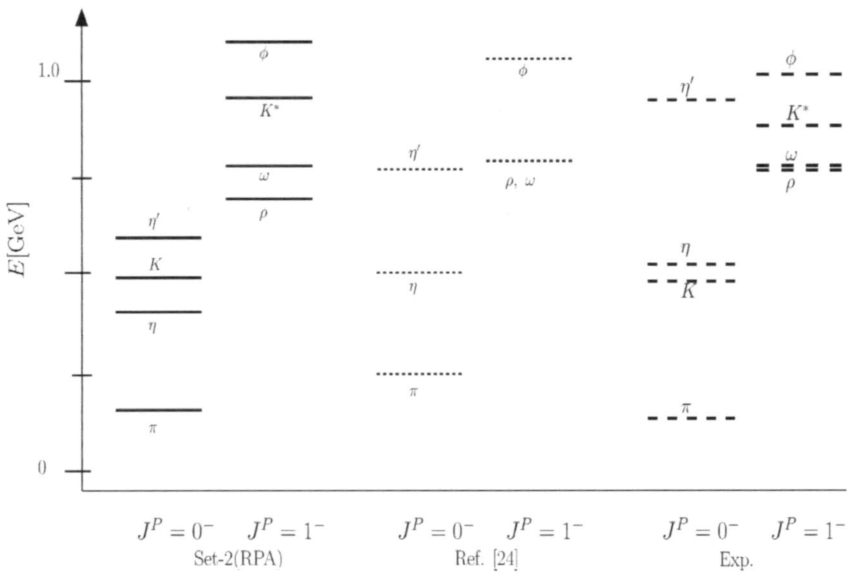

Figure 1. RPA results of Set-2 compared with experimental data and the results given in Ref. [24].

In Figure 1, we show the RPA results for Set-2 and compare them with experimental values and the results of similar approaches that use free trial wave functions.

5. Conclusions and Discussion.

We have presented a motivated QCD Hamiltonian which considers a realistic confining interaction, and used the harmonic oscillator as a trial wave functions to describe the extension of the fermion fields. We presented the exact diagonalization of the kinetic energy which provides the effective basis used for the implementation of the TDA and RPA many body methods for the diagonalization of the motivated QCD Hamiltonian.

Noticing that the $-\frac{a}{|\mathbf{x}-\mathbf{y}|}$ contribution does not modify sensibly the results at low energy, we use only the linear interaction $b|\mathbf{x}-\mathbf{y}|$ in order to investigate the low energy spectrum of pseudoscalar and vector mesons. However, the sample calculations presented here correspond to a low energy cut-off. The general case with an arbitrary cut-off, requires the renormalization of the quark masses and couplings of the interactions [22].

The TDA and RPA results indicate a pattern which resembles several features of the low energy meson spectrum but they also indicate that certain interactions, which may be related with the gluon dynamics, are missing for an even better description of the spectrum.

Acknowledgements

T. Y-M and O. C. thank to CONICET for financial support. D.A. A-Q and P. O. H thank to CONACyT (project No. 251817) and PAPIIT-DGAPA (IN100315) for financial support.

References

[1] G.S. Bali *et al* (UKQCD Collaboration), Phys. Lett. B **309**, 378 (1993).
[2] Colin J. Morningstar and Mike Peardon, Phys. Rev. D **82**, 034507 (2010).
[3] W. Bietenholz, Int. J. Mod. Phys. E **25**, 1642008 (2016).
[4] J. Dudek, Phys. Rev. D**84**, 074023 (2011).
[5] Craig D. Roberts and Anthony G. Williams, Prog. Part. NucL Phys., **33**, 477 (1994)

IOP Conf. Series: Journal of Physics: Conf. Series **876** (2017) 012022 doi:10.1088/1742-6596/876/1/012022

[6] Reinhard Alkofer and Lorenz von Smekal Phys. Rep. **353**, 281465 (2001).
[7] M. S. Bhagwat, M. A. Pichowsky, C. D. Roberts and P. C. Tandy, Phys. Rev. C **68**, 015203 (2003).
[8] A. Chodos, R. L. Jaffe, K. Johnson y C. B. Thorn, Phys. Rev. D, 10, 2599, (1974).
[9] T. DeGrand, R. L. Jaffe, K. Johndon and J. Kiskis, Phys. Rev. D, 12, 2060, (1975).
[10] A. Szczepaniak, E. S. Swanson, C-R. Ji, and S. R. Cotanch, Phys. Rev. Lett. **76**, 2011 (1996).
[11] Peter O. Hess and Adam P. Szczepaniak, Phys. Rev. C **73**, 025201 (2006).
[12] T. D. Lee, *Particle Physics and Introduction to Field Theory*, (World Scientific, Singapue, 1981).
[13] K.J. Juge, J. Kuti, and C.J. Morningstar, Nucl. Phys. B (Proc. Suppl.) **63**, 326 (1998).
[14] A. P. Szczepaniak and E. S. Swanson, Phys. Rev. D **65**, 025012 (2001).
[15] C. Feuchter and H. Reinhardt, Phys. Rev. D **70**, 105021 (2004).
[16] D. Epple, H. Reinhardt, and W. Schleifenbaum, Phys. Rev. D **75**, 045011 (2007).
[17] T. Yépez-Martínez, P. O. Hess, A. P. Szczepaniak and O. Civitarese, Phys. Rev. C**81**, 045204 (2010).
[18] T. Yépez-Martínez *et al*, Int. J. Mod. Phys. E**20**, 192199 (2011) .
[19] T. Yepez-Martinez, D. A. Amor Quiroz, P. O. Hess and O. Civitarese, J. Phys. Conf. Ser. 730, 012020 (2016).
[20] T. Yepez-Martinez, O. Civitarese, P. O. Hess Int. J. Mod. Phys. E **25**, 1650067 (2016).
[21] T. Yepez-Martinez, O. Civitarese, P. O. Hess Int. J. Mod. Phys. E **26**, 1750012 (2017).
[22] A. Amor-Quiroz, P. O. Hess, T. Yépez-Martínez, O. Civitarese and A. Weber (to be published).
[23] P. Ring, P. Schuck, *The Nuclear Many Body Problem*, (Springer, Heidelberg, 1980).
[24] F. J. Llanes-Estrada and S. R. Cotanch, Phys. Rev. Lett. **84**, 1102 (2000).
[25] C. Patrignani et al. (Particle Data Group), Chin. Phys. C, **40**, 100001 (2016).
[26] M. Moshinsky, Nucl. Phys. **13**, 104 (1959).
[27] G.P. Kamuntavicius *et al*, Nucl. Phys. A **695**, 191 (2001).
[28] J. Escher and J. P. Draayer, J. Math. Phys. **39**, 5123 (1998).

IOP Conf. Series: Journal of Physics: Conf. Series **876** (2017) 012023 doi:10.1088/1742-6596/876/1/012023

Muon study for gamma/hadron air-shower discrimination in the HAWC observatory

A. Zuñiga Reyes [1], A. Hernández Almada [1,2] and A. Sandoval Espinosa [1] for the HAWC Collaboration.

[1] Instituto de Física, Universidad Nacional Autónoma de México, Apartado Postal 20-364, México D.F 01000, México.

[2] Facultad de Ingeniería, Universidad Autónoma de Querétaro, Centro Universitario, Cerro de las Campanas, 76010, Santiago de Querétaro, México.

E-mail: asandoval@fisica.unam.mx

Abstract. The High-Altitude Water Cherenkov observatory is a ground-based array designed to study energetic gamma-rays. Experiments with this purpose have to face a huge rate of undesired hadronic background. Motivated by the fact that muon content is quite different in gamma-induced (poor in muons) and hadronic-induced (rich in muons) air-showers, we study the idea of formulating a new variable for background reduction related with counting the number of muons candidates present in such showers. Therefore, in this work we used the time differences between photomultipliers tubes to identify the signature of muons inside the water Cherenkov detectors. Showers with a high presence of muons typically must produce a characteristic time difference around 5 ns among the central detector (PMT_C) and each one of the laterals (PMT: A, B, D).

1. Introduction

Gamma-rays are the most energetic radiation of the electromagnetic spectrum. There are extreme environments in the Universe (e.g. Supernova Remnants, Active Galactic Nuclei, Gamma-Ray Bursts) which emit gamma-rays with energies in the range of GeV-TeV (Very High Energy-VHE domain). These gamma-photons are the result of electromagnetic or hadronic processes and can travel from the source to the Earth without being deflected by cosmic magnetic fields, therefore making them suitable for performing astronomy. Because the Earth´s atmosphere is opaque to photons beyond the optical waveband, VHE gamma-rays can be indirectly detected with ground-based telescopes. When a gamma-ray interact high in the atmosphere it produces a cascade of secondary relativistic particles known as an Extensive Air Shower (EAS). An air-shower is accompanied by radiation emission like: Cherenkov light, fluorescence and radio bursts. Observatories focus mainly in gamma-induced air-showers, make use of the Cherenkov effect in both: air and water. Imaging Atmospheric Cherenkov Telescopes (IACTs) collect this light with mirrors and reflect it toward sensitive cameras, producing as a result an image of the shower. Depending of the image´s shape it can be identified as gamma-initiated showers and also to estimate their energy and direction. Other technique, employed by surface array telescopes, consist in detecting the light generated when charged particles of the shower

IOP Conf. Series: Journal of Physics: Conf. Series **876** (2017) 012023 doi:10.1088/1742-6596/876/1/012023

enter inside water reservoirs. A common issue of these experiments is dealing with a large background originated by hadronic primary cosmic rays. Such "noise" can be thousands of times greater than gamma signals (in HAWC is ~ 24 kHz [1]), then new methods for gamma/hadron separations are needed. The muon content of the showers could be a good discrimination parameter, because gamma showers are purely electromagnetic and contain few muons. In order to help the noise reduction in HAWC, this work aim on identifying muons signals in data using only times measurements.

2. The High-Altitude Water Cherenkov observatory

The High-Altitude Water Cherenkov (HAWC) observatory is a facility conceived to study gamma-rays with energies between 100 GeV and 100 TeV. It is located 4100 m a.s.l on the flanks of the Sierra Negra volcano near Puebla, Mexico. The detector is an array of 300 Water Cherenkov Detectors (WCD) or tanks, each one with 4.5 m high, 7.3 m diameter and a volume that contains 200 000 L of purified water. There are four photomultiplier tubes (PMTs) positioned in the tank base: a high-quantum efficiency detector or PMT_C (10-inch diameter) centrally located, surrounded by three (8-inch diameter) PMTs (A, B, D); all upward facing (Fig. 1). With an instantaneous field of view of 2 sr, HAWC observes two-thirds of the sky during each 24-hour period in the Northern Hemisphere [2].

2.1. HAWC operating principle

The HAWC array is used to sample air-shower particles by recording the Cherenkov light produced when they pass through tanks full of water. This electromagnetic radiation is emitted into a forward cone that surrounds the direction of motion of the charged particle. The opening angle of the cone depends on the index of refraction (n) of the medium, in water (where n = 1.33) the angle is ≈ 41°. Because the Cherenkov cone in water is so large, nearly every charged particle that enters the tank should be observed by at least one of the four PMTs, which convert the light into an electrical signal measurable by the data acquisition system. The output data correspond to charge and time of triggered PMTs, later used for the shower´s parameters reconstruction (e.g. core position, arrival direction, energy, event classification: gamma or hadron).

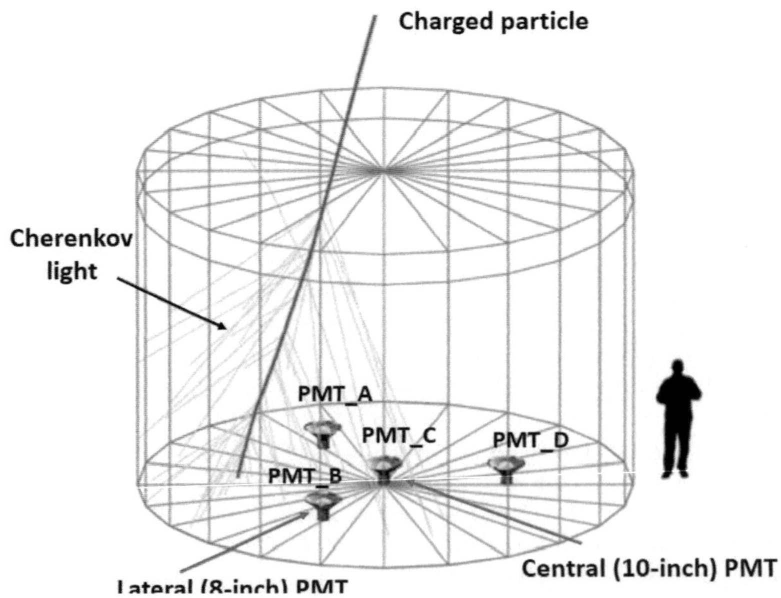

Figure 1. WCD setup, where the passage of a charged particle emitting Cherenkov photons is illustrated.

3. Muon candidates in HAWC raw data

Muons are the most abundant charged particle at the HAWC altitude and arrive mainly with some GeV of energy. They are originated from charged meson decays (π, K) in EAS, following approximately straight trajectories through the atmosphere and even inside WCD. Depending on the area where a muon travels it can trigger a specific number of PMTs, this is called multiplicity (M).
In this work we used a sample of raw data. This refers to the processed signal of PMTs before the multiplicity trigger threshold for events selection is applied. In order to detect muon candidates in raw data, we analysed tanks with four-folds (M = 4) in a very short time window. This selection can include electrons which trigger as well all 4-PMTs inside a WCD, but there is a specific difference in relation to the time of detection they produce.

3.1. Simulations to compare with selected data

Regarding vertical muons with trajectories close to the central light-sensor, in [3] it was analytically calculated a characteristic time difference (≈ 5.4 ns) between the PMT_C relative to the peripheral ones. With the aim of comparing the time differences distributions obtained in raw data sample, we performed Monte Carlo simulations (MCsim) of vertical muons and electrons in a single tank (Fig. 2).

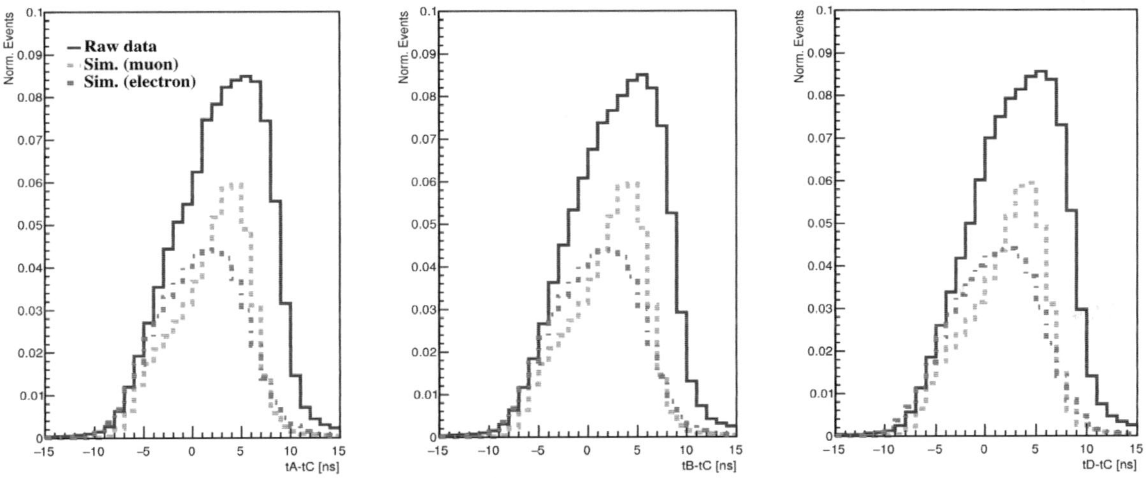

Figure 2. Comparison of the time difference distributions between PMTs for vertical electrons (red) / muons (green) and raw data (blue).

In the simulation we set the following energy for particles: electrons with 1 - 500 MeV and muons with 15 GeV. Data selection was in tanks with M = 4 within 15 ns window. Fig. 2 show that raw data and vertical muon distributions present a similar shape, proving that time difference (PMT_laterals – PMT_C) has a mean value of approximately 5 ns. However, electron distribution is quite different with a time difference average value near 0 ns, meaning that the PMTs see the Cherenkov light practically at the same time. This happens because electrons stop quickly in the first centimetres of the tank and then the light travels to the bottom falling promptly over all the detectors. On the other hand, muons reach the tank floor triggering the PMT_C first, and later the light arrive to the lateral PMTs.

4. Time difference analysis in shower data

Like gamma showers are mainly electromagnetic (containing a lot of e-, e+ and γ), they must exhibit a time difference close to 0 ns. In order to test this hypothesis, we used MCsim data of gamma-induced showers created by the HAWC software and real shower data which passed the trigger condition.

Figure 3 shows distributions produced by showers with unequal muon content. The shower data is rich in muons, this is the reason why the time differences tends to approximately 5 ns.

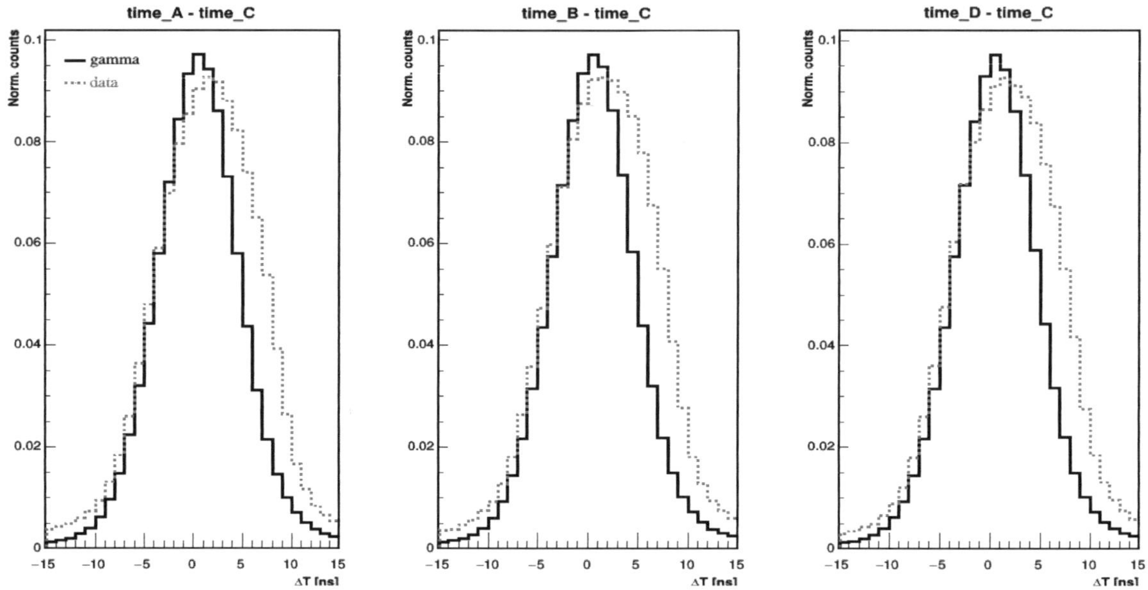

Figure 3. Comparison, using time difference distributions, of shower data (red) vs gamma MCsim (black) for tanks with multiplicity 4.

An overlap of both data can be seen in Fig. 4, that was created using two time differences. It is evident the spreading of shower data toward large time values. Each point represents a tank with M = 4. Figure 5 include charge information for a pair of PMTs used in a time difference. Most of the values are displaced to left side as a consequence of central PMT higher sensitivity.

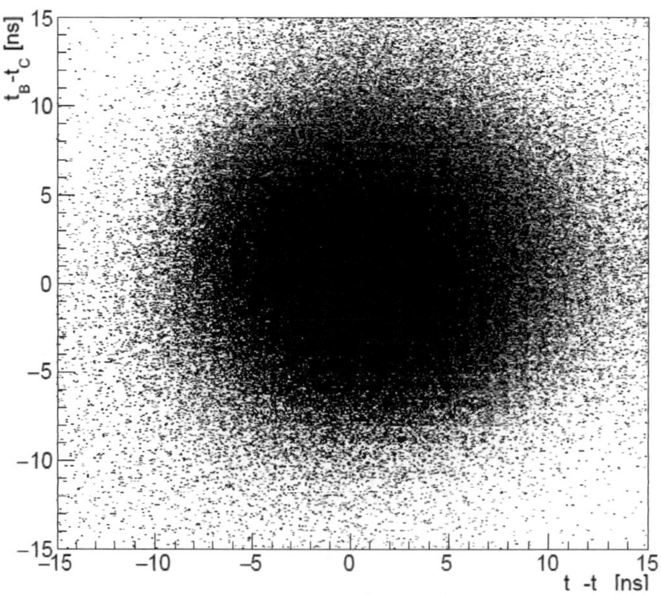

Figure 4. 2D plot with two time differences corresponding to gamma MCsim data (black points) and shower data (red points).

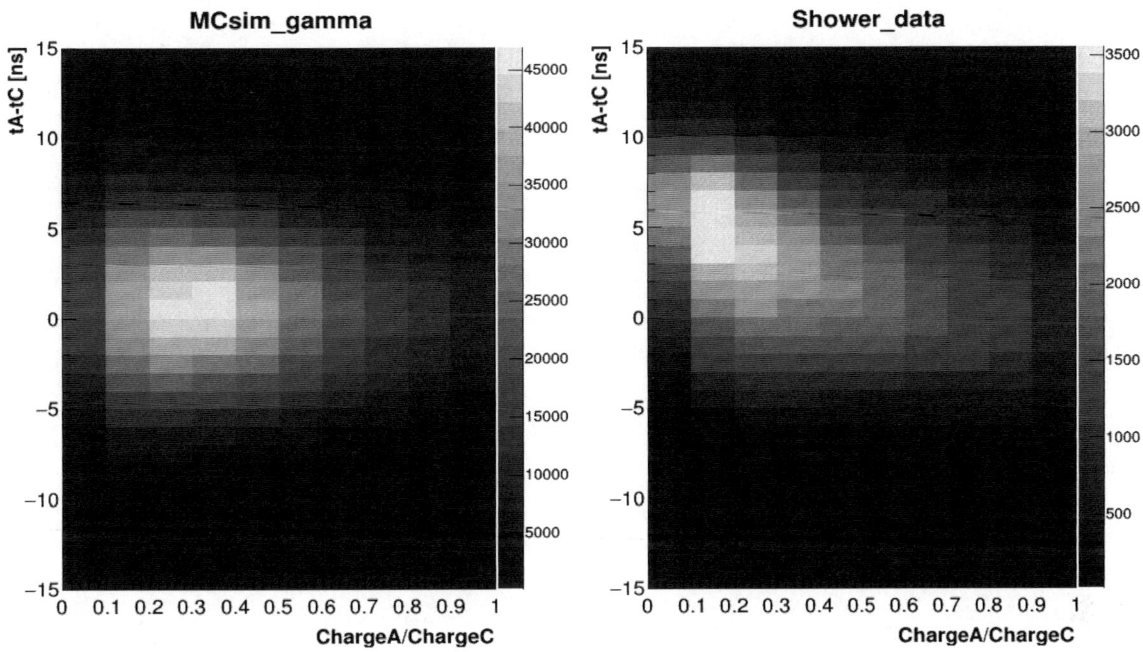

Figure 5. Comparison of 2D plots built with a time difference and the corresponding charge ratio, for gamma MCsim and shower data.

5. Remarks and future work

- Muon candidates are detected in HAWC data with a characteristic time difference close to 5 ns which is different for electrons.
- There is a separation between the time difference distributions for gamma MCsim and the shower data, due to the unequal muons content.

The future work is to create a discrimination variable that consider the number of muons in a shower. One muon candidate will be selected using the restriction r_time > 7 ns, where r_time is presented in equation (1). Such cut allows to obtain a higher discrimination power. Then, the number of muon candidates will be the sum of tanks with multiplicity 4 per event that satisfied the restriction, expecting larger number for hadronic showers.

$$r\_time = \sqrt{(tA - tC)^2 + (tB - tC)^2 + (tD - tC)^2} \quad (1)$$

Acknowledgments

We acknowledge the support from: US National Science Foundation (NSF); US Department of Energy Office of High-Energy Physics; The Laboratory Directed Research and Development (LDRD) program of Los Alamos National Laboratory; Consejo Nacional de Ciencia y Tecnología (CONACyT), México; Red de Física de Altas Energías, México; and the University of Wisconsin Alumni Research Foundation.

References

[1] Abeysekara A U *et. al.* 2017 ApJ arXiv:1701.01778v1 [astro-ph.HE]
[2] Smith A J 2015 ICRC arXiv:1508.05826v2 [astro-ph.IM]
[3] Wood J [HAWC Collaboration internal note] 2015